COMMONLY ASKED QUESTIONS IN
PHYSICS

COMMONLY ASKED QUESTIONS IN

PHYSICS

ANDREW REX

CRC Press
Taylor & Francis Group
Boca Raton London New York

CRC Press is an imprint of the
Taylor & Francis Group, an **informa** business

First published 2014 by CRC Press

Published 2019 by CRC Press
Taylor & Francis Group
6000 Broken Sound Parkway NW, Suite 300
Boca Raton, FL 33487-2742

© 2014 by Taylor & Francis Group, LLC
CRC Press is an imprint of the Taylor & Francis Group, an informa business

No claim to original U.S. Government works

ISBN-13: 978-1-4665-6017-8 (pbk)

Visit the Taylor & Francis Web site at
http://www.taylorandfrancis.com

and the CRC Press Web site at
http://www.crcpress.com

Contents

Preface

When the editors at Taylor & Francis contacted me to suggest this project, I was skeptical at first. I've written college-level textbooks but had never before attempted a book intended for such a wide audience. As we began to discuss the project, however, I grew more excited about the possibilities. Those of us who practice physics for a living frequently encounter people from all walks of life who, once they know our profession, pepper us with questions about anything and everything that smacks of physics. In our modern world with its rapid electronic communication, you often hear brief sound bites that mention the latest discoveries, such as the Higgs boson, dark energy, or fusion-powered energy. When they hear or read about these, people naturally want to find out more to fill in the gaps.

I use that word "naturally" quite intentionally. Part of what makes us human is our interest in the world around us. For quite a lot of human history, we stumbled in the dark trying to understand the forces of nature. The scientific revolution of Galileo and Newton gave us not only the field we recognize as physics, but also the practice that has grown into our modern scientific method. In the 300 years since Newton, physics has explained many things that used to be mysterious. Particularly in the last century, physics has addressed a wider range of questions, from the smallest fundamental particles to the large-scale structure and history of the entire universe.

But there are always more questions. One thing leads to another—that's the path of science. There are so many questions that there's no way a single book like this one can cover them all. However, this book does contain many of the most commonly asked questions, as the editors and I have determined them from our surveys of physics students and the wider public. The book's content covers a wide range of subjects, from older physics that goes back to the age of Newton to new ideas only formulated in the twenty-first century. There are chapters devoted to core areas of physics that predate the twentieth century: mechanics (Chapter 1), electromagnetism (Chapter 2), optics (Chapter 5), and thermodynamics (Chapter 6). However, because there is such intense curiosity about modern physics, I decided to place significant focus on that area. This includes quantum mechanics (Chapter 4), atomic and nuclear physics (Chapter 7), fundamental particles (Chapter 8), and relativity (Chapter 9).

Because this is not a textbook and because much of the intended audience doesn't have a background in advanced mathematics, I have scaled back the mathematics level from what one might find in even an introductory-level physics text. In every chapter there are discussions of numbers and the units we use to measure things, because measurement and experimental work allow us to say that we know what we know, and the precision of measurement gives us confidence in the results. Some chapters include a "Going Deeper" feature that provides more mathematical details for readers who feel up to the challenge. Others can skip those boxed areas and move on to the next question. The suggested readings at the end of each chapter range from classic textbooks to some of the best books written for the general public, in case the questions in that chapter have made you want to study that topic in more depth.

There should be something here for everyone. My own students, like other students from around the country and around the world, come to college with a lot of ideas and questions about physics. This book should be a great resource for them, whether or not they pursue physics as a major. There's a much larger audience outside our colleges and universities. In our increasingly technical world, there are many people whose professional work is touched by developments in physics. This includes medical practitioners, scientists in other fields, engineers, and teachers. And if you've picked up this book to look at it, you probably realize that physics affects your life nearly every day. (Have you ever had an x-ray or used a cell phone?) This book is written for you, for all of you who are curious and want to know more about your world and universe.

Acknowledgments

I want to express my appreciation to Luna Han, Taylor & Francis editor, who has provided encouragement and useful suggestions at every stage of this project. My students Audrey Kvam and Kyle Whitcomb read manuscript chapters and offered their valuable perspectives as curious young learners of physics. My colleague Bernie Bates, who also read part of the manuscript, offered his thoughtful professional analysis. And last but not least, I am fortunate to have the support of my family—Sharon and Jessica—who have encouraged me on this project, as always.

About the Author

Andrew Rex is professor of physics at the University of Puget Sound in Tacoma, Washington. He received the BA in physics from Illinois Wesleyan University in 1977 and the PhD in physics from the University of Virginia in 1982. At Virginia he worked under the direction of Bascom S. Deaver, Jr., on the development of new superconducting materials. After completing requirements for the PhD, he joined the faculty at Puget Sound.

Dr. Rex's primary research interest is in the foundations of the second law of thermodynamics. He has published research articles and, jointly with Harvey Leff, two comprehensive monographs on the subject of Maxwell's demon (1990, 2003). Dr. Rex has coauthored several widely used textbooks: *Modern Physics for Scientists and Engineers* (1993, 2000, 2006, 2013), *Integrated Physics and Calculus* (in two volumes, 2000), and *Essential College Physics* (in two volumes, 2010).

Dr. Rex has served in administrative roles, including as chair of his department and director of the University of Puget Sound Honors Program. He is devoted to physics education and has been an active participant in the American Association of Physics Teachers, the Society of Physics Students, Sigma Pi Sigma, and Sigma Xi. In 2004 Dr. Rex was recognized for his teaching with the president's award for teaching excellence.

Chapter 1

Classical Mechanics

All around us we see bodies in motion, and a big part of physics concerns how and why things move the way they do. Quite a lot of what you see can be explained by **classical mechanics,** a branch of physics that goes back to the seventeenth century. Many of the basic concepts and ideas came from Isaac Newton (1642–1727), so physicists often use the term **Newtonian mechanics** interchangeably with classical mechanics. Since the days of Newton, the field has been revised and expanded to include new concepts, particularly energy, which along with better computational methods has improved the power of classical mechanics to explain what we see and to make predictions about how physical systems will behave. Perhaps surprisingly, much of the world around you can be explained using relatively few simple rules and concepts.

WHAT IS PHYSICS?

Classical mechanics is just one part of the much larger subject that we call physics. Broadly speaking, physics is the study of the whole universe around us, ranging from the largest structures—galaxies and groups of galaxies—to the smallest subatomic particles, and everything in between. One concept central to physics is the study of forces, which govern the interactions of objects and particles with one another. Another key concept is energy, which allows us to analyze many key processes and transformations. Both energy and force are useful in studying how things move in space and time.

Many of the key concepts in physics, including force and energy, had their roots in early classical mechanics. In that context, they're still useful today to help us understand the motion of some everyday things, from falling balls to simple machines. However, these old concepts are now applied in ways that go far beyond classical mechanics. For example, energy is an essential concept in quantum mechanics, where new approaches have helped us to understand

1

processes in atoms and nuclei that classical mechanics can't explain. And in the last century, physicists have identified new forces (called the strong and weak forces) that act only between subatomic particles over extremely short ranges. The number of topics and applications included in the study of physics has grown considerably since the time of Newton, but we still find classical mechanics useful in a variety of settings.

 WHAT IS THE SI SYSTEM OF UNITS?

Sometimes people associate physics only with big ideas—gravity, nuclear energy, atoms and particles, and so on. But none of these ideas would be meaningful if we couldn't measure and quantify the phenomena we see. Scientists generally use the SI system of units (from the French *système international d'unités*) for measurement. The SI system gives us a sensible common language for recoding measurements and for doing computations.

The three SI units that serve as the basis for classical mechanics are the meter (for length), kilogram (for mass), and second (for time). Measurements and computations are normally reported using the abbreviations m, kg, and s, respectively, as in a length of 3.2 m or a mass of 56 kg.

To avoid long strings of zeroes at the end of a number or after a decimal point, we use **scientific notation,** in which a measured or computed quantity is expressed as a number multiplied by a power of 10. For example, Earth's mass is about 5.97×10^{24} kg, and the electron's mass is 9.11×10^{-31} kg. By convention, quantities are normally reported with a single number to the left of the decimal point.

SI prefixes can be used as an alternative to powers of 10. Most people are familiar with using centimeters (cm) and millimeters (mm) for measuring length with a ruler. Table 1.1 shows some other SI prefixes. This table contains prefixes for a wide range of numbers but is not exhaustive. Note that many prefixes occur at intervals of $10^3 = 1,000$ with respect to the next closest prefixes, larger and smaller. That way, you can express physical quantities using an appropriate prefix with a three-digit number or smaller. For example, a time interval of 4.5×10^{-5} s is 45 μs, and a distance of 8.25×10^5 m is 825 km. For the most part SI prefixes and powers of 10 are interchangeable, but in some cases one style is preferred by convention. For example, physicists usually write 633 nm rather than 6.33×10^{-7} m for the wavelength of red light from a helium-neon laser.

How Are SI Units Defined?

SI units are defined precisely and by international agreement, so that there will be a single set of standard units used worldwide. You might think that a second should be defined as some fraction of a year. However, the length of the

TABLE 1.1 SOME SI PREFIXES

Power of 10	Prefix	Symbol
10^{-15}	Femto	f
10^{-12}	Pico	p
10^{-9}	Nano	n
10^{-6}	Micro	μ
10^{-3}	Milli	m
10^{-2}	Centi	c
10^{-1}	Deci	d
10^{3}	Kilo	k
10^{6}	Mega	M
10^{9}	Giga	G
10^{12}	Tera	T
10^{15}	Peta	P

year doesn't stay constant, due to changes in Earth's orbit. Instead, a second is defined as 9,192,631,770 periods of the radiation from a transition between two energy levels in a ^{133}Cs atom. This atomic standard is highly reproducible and reasonably accessible for those who wish to use it. Similarly, the meter is defined as the distance traveled by light in 1/299,792,458 s. This modern definition of the meter replaced the practice of using the length of a single metal bar, which could not be accessed universally and was subject to thermal expansion and other changes in time. Defining the meter with light makes it more accessible to all, and it's convenient because the speed of light—a universal constant—has been redefined as a nine-digit quantity: 299,792,458 m/s. The kilogram is still defined using an artifact, a cylinder of platinum–iridium alloy kept in the International Bureau of Weights and Measures in France. However, this artifact may soon be replaced by an electronic standard based on Planck's constant (see Chapter 4), which, like the speed of light, is a fundamental constant of nature.

What Are SI Base Units and Derived Units?

Having independent definitions makes the second, meter, and kilogram **base units** in the SI system. There are four other base units: ampere (A) for electric current, kelvin (K) for temperature, candela (cd) for luminous intensity, and mole (mol) for the amount of a substance. Notice that SI units written out are not capitalized, even when named for a person (e.g., kelvin or newton). **Derived units** are defined in terms of base units. An example of a derived unit in classical mechanics is the unit for energy, which in terms of base units is kg·m^2/s^2. The derived unit is the joule (J), defined as 1 J = 1 kg·m^2/s^2. It's convenient to use derived units whenever you can, to save the trouble of writing what can be a complex string of base units.

The one-to-one conversion factor, as in 1 J = 1 kg·m^2/s^2, is a hallmark of the SI system, with other derived units defined in a similar way. Having simple conversions is one of the two chief advantages of the system, with the other being the universal accessibility of standards. This is just what the creators of SI had in mind when they developed it in France in the late eighteenth century. They first defined the meter as one ten-millionth of the arc along Earth's surface from the North Pole to equator. (Today's meter is not far off from that definition!) With the meter in hand, they defined the gram as the mass of 1 cubic centimeter of water, so that water's density would be exactly 1 g/cm^3 or 1000 kg/m^3. Conversions would then follow as in today's system, all based on powers of 10. This replaced the old English system (e.g., 12 inches = 1 foot and 5280 feet = 1 mile), and a similarly haphazard set of French conversions. Although the base unit definitions have been updated, the spirit of simplicity and accessibility remains.

Are SI Units the Only Ones Used in Physics?

Despite the advantages of the SI system, physicists sometimes find it sensible to use non-SI units for their work. Generally, this is done only when a different unit has its own advantages. One example is the **atomic mass unit** (u), used to measure very small masses, particularly those in atoms and nuclei. 1 u is defined to be 1/12 of the mass of the ^{12}C atom. Thus, that atom has a mass of exactly 12 u. Other atoms have masses very close to a whole number of atomic mass units (e.g., 4.003 u for ^{4}He and 23.985 u for ^{24}Mg). The proton (1.008 u) and neutron (1.009 u) are the fundamental building blocks of all nuclei, and their masses are conveniently close to 1 u each. In this case the alternative of SI units has no such integer patterns and requires all masses to have large negative exponents (such as 10^{-27} kg).

Another example on a much larger scale is the **light year** (ly), which is defined as the distance light travels in 1 year and is about 9.46 × 10^{15} m, an inconveniently large number. You can express the distance from the sun to its nearest neighbor star as just over 4 ly, and many other stars are less than 100 ly away. On that scale of distances, the light year is easier to use than the meter.

WHAT ARE VELOCITY AND ACCELERATION?

Velocity and **acceleration** are quantities used to describe an object's motion. Both are based on knowing the object's position as a function of time. Velocity, measured in m/s, is the rate of change of position, and acceleration, measured in meters per second squared, is the rate of change of velocity.

In one-dimensional motion (think of a car driving along a straight road), position is given by a single number x, measured in meters (m) on a coordinate axis. Suppose the car travels at a constant rate from $x = 20$ m to $x = 100$ m in a time of 4.0 s. Then its constant velocity is 80 m/4.0 s = 20 m/s. Another car traveling at the same constant rate in the opposite direction has a velocity of –20 m/s, with the negative sign indicating the opposite direction of travel. Acceleration is any change in velocity. If the car traveling at 20 m/s starts going faster, its acceleration is positive, and if it slows down, its acceleration is negative.

Motion in two or three dimensions requires that position, velocity, and acceleration be expressed as **vector** quantities. A vector is an ordered set of numbers—two numbers for two-dimensional motion and three numbers for three-dimensional motion. In two dimensions, the ordered pair of **components** (x,y) is used to measure position, relative to perpendicular coordinate axes x and y. The velocity vector has components v_x and v_y that are the rates of change of the x and y coordinates, respectively, of the body in motion. For example, a velocity vector (8.2 m/s, –4.1 m/s) describes an object moving simultaneously in the +x-direction and –y-direction, with the rate of motion in the +x-direction twice the other rate. Similarly, acceleration components a_x and a_y, which are the rates of change of v_x and v_y, and each acceleration component can be positive or negative, independently of the other one.

GOING DEEPER—VELOCITY AND ACCELERATION WITH CALCULUS

Velocity and acceleration are both defined as rates of change. In calculus, the **derivative** is also a rate of change, so it's the perfect tool for relating position, velocity, and acceleration.

In one-dimensional motion (along the x-axis), **average velocity** over a time interval Δt is

$$v_{av} = \frac{\Delta x}{\Delta t} \tag{1.1}$$

Velocity v at any single moment in the interval is found by taking the limit of the average velocity as the time interval shrinks to zero:

$$v = \lim_{\Delta t \to 0} \frac{\Delta x}{\Delta t} \tag{1.2}$$

The velocity v given by Equation (1.2) is sometimes called the **instan-taneous velocity** because it refers to the velocity at one instant of time. This matches the definition of derivative from calculus, so you can say that the velocity is simply the derivative of the position with respect to time. Symbolically,

$$v = \frac{dx}{dt} \tag{1.3}$$

Equation (1.3) can be taken as the exact definition of velocity for one-dimensional motion. With this definition, the physical units for velocity come out automatically. With position in meters and time in seconds, velocity is in meters per second. For example, a car traveling down a highway might have a position function $x = 25t$ m. Then its velocity is $v = dx/dt = 25$ m/s.

Similarly, acceleration is the derivative of velocity with respect to time:

$$a = \frac{dv}{dt} \tag{1.4}$$

In two or three dimensions, velocity is a vector, with each component given by the derivative of the appropriate position component:

$$v_x = \frac{dx}{dt} \quad v_y = \frac{dy}{dt} \quad v_z = \frac{dz}{dt} \tag{1.5}$$

Acceleration is the rate of change of velocity, so it's also given by derivatives:

$$a_x = \frac{dv_x}{dt} \quad a_y = \frac{dv_y}{dt} \quad a_z = \frac{dv_z}{dt} \tag{1.6}$$

Are Velocity and Speed the Same?

In everyday language, people sometimes use velocity and speed interchange-ably, but in physics they're not the same. In one-dimensional motion, veloc-ity is the rate of change of position and is either positive or negative, depending on the direction of travel relative to a defined coordinate axis. **Speed** in one-dimensional motion is the absolute value of the velocity. Thus, speed is always a positive number (or zero, for an object at rest), regardless of the direction of motion, with the same units as velocity—meters per second in the SI system. Your car's "speedometer" really does measure speed, not velocity, because the single number indicates your rate of travel without regard to direction.

In two or three dimensions, velocity is a vector, made up of two or three components that give information about the rate of travel along each coordinate axis. Speed is a single number that's equal to the velocity vector's **magnitude.** To understand what the magnitude of a vector means, think about two points in a plane (Figure 1.1). The distance d between the two points (x_1, y_1) and (x_2, y_2) is given geometrically as

$$d = \sqrt{\Delta x^2 + \Delta y^2}$$

By similar reasoning, speed in two dimensions is computed from the velocity components v_x and v_y:

$$v = \sqrt{v_x^2 + v_y^2} \tag{1.7}$$

Thus, just as in one dimension, speed is a single number with units (m/s) in SI, and it must be either positive or zero. Physically, speed tells you the rate of travel without reference to the direction of travel.

An alternative way to describe velocity in two or three dimensions is to give the speed and the direction of travel. This is fairly straightforward in two

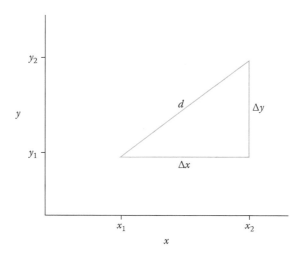

Figure 1.1   The distance between two points in a plane.

dimensions, where direction can be specified by a single number. For example, in polar coordinates, the direction is given as an angle with respect to a fixed axis. In three dimensions it's a little harder to specify direction, but it can be done with two angles, as in spherical polar coordinates.

 ## WHAT IS A FORCE?

Some obvious examples of **forces** are when you push or pull on something. When you drop an object, gravity is the force that pulls it toward Earth. A direct "push–pull" involves direct contact between two objects, but gravity doesn't require contact, so it's called an "action at a distance" force. Other familiar forces that act through a distance are electricity and magnetism. What all forces have in common is that they involve an interaction between two objects. Further, force is a vector because each force has a particular strength and a specific direction.

Dynamics is the study of forces and the changes in motion. **Kinematics** is the study of motion alone—position velocity, and acceleration—without regard to the forces that cause changes in motion.

What Are Newton's Laws of Motion?

Newton's three laws of motion provide the basis for understanding how objects behave under the influence of forces. The first two involve the concept of **net force,** which is simply the sum of all the forces acting on an object.

Newton's first law: If the net force on an object is zero, then its velocity is constant.

Note that "constant velocity" can mean either moving with a constant speed in a fixed direction or not moving at all (i.e., zero velocity). These two cases—moving or not moving—are mentioned explicitly in a more colloquial form of Newton's first law: An object at rest remains at rest, and an object in motion continues to move with constant velocity, unless acted upon by a net force.

Newton's second law: An object's acceleration is directly proportional to the net force acting upon it:

$$F_{net} = ma \qquad (1.8)$$

where m is the object's mass. Mass is measured in kilograms. You can think of mass as a quantity of matter or, in the spirit of Equation (1.8), as resistance to motion. That's because $a = F_{net}/m$, so for a given force, the acceleration varies inversely with mass—more mass means less acceleration.

Force and acceleration are both vectors, but mass is not. Mass is always positive, so the vectors F_{net} and a are in the same direction. That is, an object's acceleration is always in the direction of the net force. This should make intuitive sense. To make an object at rest move in some desired direction, you push or pull in that direction.

Equation (1.8) helps define the units for force. With mass in kilograms and acceleration in meters per second squared, the units for force are kg·m/s². That combination of units is defined as the newton (N), so 1 N = 1 kg·m/s².

Newton's third law: When two objects interact, the force on one is equal in strength and opposite in direction to the force on the other.

Newton's third law says that forces always come in pairs. When you push on a wall, the wall pushes back on you with the same amount of force in the opposite direction. The law works on all forces, even ones that act at a distance. If you drop a rock, Earth pulls downward on the rock with the force of gravity, but the rock attracts the Earth upward with the same amount of force. Then why does the rock move and not Earth? Look at Newton's second law. With a mass on the order of 1 kg, the rock is more susceptible to acceleration than Earth, with its mass of 6×10^{24} kg.

There's a colloquial version of Newton's third law that goes: For every action, there's an equal and opposite reaction. The sense of this may agree with Newton's third law, but physicists don't care for the language. First, "action" is a different physical quantity from force. (We won't define it in this book.) Second, two vectors aren't really equal if they're in opposite directions. The statement of Newton's third law given previously is preferred.

Are Mass and Weight the Same?

Mass and weight are very different things. Mass is an amount of matter or resistance to force, measured in kilograms. Weight is a force, measured in newtons. Specifically, **weight** is the force of gravity. An object's mass depends only on the matter it's composed of and is the same anywhere in the universe. An object's weight depends on where it is relative to other bodies and represents the force of gravity due to those bodies.

The concept of weight is most applicable if you're near the surface of a large body, such as Earth. There the weight you feel is due to your gravitational attraction toward Earth. If you were on the moon, your mass would be the same but your weight only about one-sixth as much as on Earth. If you were in outer space, far from Earth, moon, or other massive bodies, your weight would be essentially zero. We say "essentially" and not exactly zero because you'll always feel some attraction to other masses in the universe. Remember: Newton's second law deals with the net force due to all the forces acting on you, and that's true for weight as for any other force.

Why Do Different Masses (or Weights) Fall at the Same Rate?

Galileo is reported to have dropped two balls having different masses, in about 1589, to refute the claim that heavier (or more massive) balls should fall faster than lighter ones. Galileo claimed that the two balls should accelerate downward at the same rate, due to Earth's gravity. Newton's laws of motion explain why this is so. For any object with mass m and weight W in free fall, its downward acceleration is $g = W/m$, according to Newton's second law. But weight is proportional to mass, so for any object the ratio W/m is the same, making the downward acceleration g the same for all. Near Earth's surface g is about 9.8 m/s^2, though this varies slightly due to altitude and latitude.

What Are Friction and Drag Forces?

Try Galileo's experiment for yourself by dropping a pencil and a piece of paper simultaneously. At first, leave the paper unfolded. You'll find that the pencil reaches the ground much faster than the paper. Don't worry, there's nothing wrong with gravity! The paper experiences a **drag force**—the force experienced by an object traveling through a fluid—because it's running into the air on its way down. The paper pushes downward on the air, and by Newton's third law, the air pushes up on the paper, so the net force on the paper (gravity plus drag force) is much smaller than gravity alone. The skinny pencil experiences much less drag force, so it wins the race to the floor. You can nearly equalize the drag forces, and accomplish Galileo's predicted outcome, if you crumple the paper into a tight ball before you drop it. This minimizes the drag force and lets the paper fall at nearly the same rate as the pencil.

To his credit, Galileo understood something about drag forces because he had experimented extensively with fluids. He noted that a heavy ball and light ball dropped from the Tower of Pisa won't fall together precisely. Rather, the one with less mass loses the race by a whisker because it's affected slightly more by drag than the more massive one, just like your paper and pencil. Take away the air, Galileo said, and they would fall together. In 1971 astronauts famously repeated this experiment by dropping a hammer and feather on the moon, where there's virtually no atmosphere. The two objects fell together, vindicating Galileo and Newton.

Drag forces can have a big effect on motion, and the amount of drag increases the faster you go. You've felt this if you've ever tried to go fast on a bicycle. Cars are designed to limit drag forces, to increase fuel efficiency. When a jet airplane is cruising at constant speed, the force from its powerful engines is just balanced by the drag force in the opposite direction, resulting in zero net force and no change in speed (Newton's first law). Figure 1.2 shows the significant effect drag can have on a ball game—in this case a batted

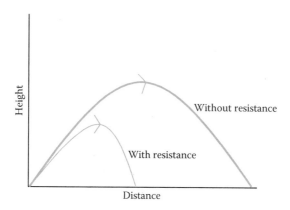

Figure 1.2   Flight of a baseball, with and without drag.

baseball. Swimming or making a boat go through water is an even more difficult task because a dense fluid like water makes for strong drag forces, even at low speeds.

Frictional forces result from the intermolecular forces that occur when one surface moves over another one. Slide a book across a desk, and the force of friction slows it and eventually stops it. The strength of the frictional force depends highly on the surfaces and how they interact with one another. A hockey puck sliding on smooth ice experiences little friction. Inside your car engine, oil reduces the frictional forces between metal surfaces. Sometimes you want to maximize frictional forces—for example, between your tires and the roadway. The simple act of walking depends on the force of friction between your feet and the ground. A short walk across a patch of ice should convince you how useful friction is in this case.

WHAT ARE WORK AND ENERGY?

Work is a measure of the effectiveness of an applied force or forces and is defined as the product of force and distance moved. If the force is not in the direction of motion, then only the component of the force in the direction of motion is used. Work has SI units N·m; this is defined as the joule (J), so 1 N·m = 1 J. Work can be positive, when the force is in the direction of motion, or negative, when it's in the opposite direction.

Energy can't be defined simply because it takes so many forms. The form most closely related to work is **kinetic energy.** An object with mass m and speed v has kinetic energy

$$K = \tfrac{1}{2}\, mv^2 \qquad\qquad (1.9)$$

The SI units for kinetic energy are kg·m^2/s^2, which is equivalent to joules (J). All forms of energy can be expressed in joules.

What Is the Work–Energy Theorem?

Because work and kinetic energy have the same units, you might guess that there's a connection between them, and there is. Applying a force and doing work change kinetic energy. If multiple forces are acting, the net work W_{net} due to all the forces is the sum of the work done by all the individual forces. The **work–energy theorem** says that

$$W_{net} = \Delta K \qquad\qquad (1.10)$$

where ΔK means the change in kinetic energy. One way in which the work–energy theorem makes sense is that positive net work results in an increase in kinetic energy, while negative net work results in a decrease in kinetic energy.

What Is Potential Energy?

By doing work, forces act to increase and decrease the kinetic energy of objects within a system. If the kinetic energy that's removed can be stored and recovered later, the stored energy is called **potential energy.** Potential energy (symbol U) is always associated with a particular force. Forces that allow kinetic energy to be stored and recovered later are called **conservative forces,** and conservative forces have potential energy associated with them. Gravity is an example of a conservative force. **Nonconservative forces,** for which lost kinetic energy can't be recovered, have no associated potential energy. Frictional and drag forces are nonconservative.

What Is Conservation of Mechanical Energy?

Total mechanical energy (E) is the sum of kinetic and potential energy in a system. If the forces acting are conservative, then total mechanical energy is conserved. That is, $E = K + U$ is a constant.

As an example, think about what happens when you throw a ball straight up. Once the ball is in flight, the force of gravity makes it slow down, eventually stopping and falling back to its starting point. During the ball's upward flight, its kinetic energy steadily decreases, while its potential energy increases. On the return path downward, the ball's kinetic energy increases while its potential energy decreases. If the drag force due to the ball's interaction with air is negligible, the total mechanical energy $E = K + U$ is constant throughout the ball's flight. Increases in kinetic energy are offset by decreases in potential energy, and vice versa.

The effect of nonconservative forces is to reduce the total mechanical energy. For the ball projected upward, the drag force isn't negligible if you throw the ball fast enough. That force is directed opposite to the ball's motion (whether it's going up or down), so it does negative work on the ball, reducing its total mechanical energy. In general, the change in mechanical energy ΔE is equal to the work done by nonconservative forces.

What Is Power?

Power is the rate at which work is done or the rate at which energy is supplied or used. The SI units for power are joules per second, which is defined to be a watt (W), so 1 W = 1 J/s.

The concept of power is the same, regardless of what kind of energy is being used. In this chapter we've defined kinetic energy and mechanical energy. In other chapters we'll address questions about electrical energy (Chapter 2), thermal energy (Chapter 6), and nuclear energy (Chapter 7).

You may be familiar with the concept of power from its use in electrical energy. Electric lights and other devices are given a rating based on the power they use, as in a 60 W light bulb or a 200 W motor. Some electric utility providers measure energy in units of kW·h. This is a non-SI unit, but it makes sense for energy delivery. If power is energy/time, then energy is power × time, so kW·h is a valid measure of energy. If you run your 60 W light bulb for 20 h, you'll use 60 W × 20 h = 1200 W·h = 1.2 kW·h of energy.

WHAT IS MOMENTUM CONSERVATION?

A particle of mass m and velocity v has a momentum (symbol p) given by mv. Momentum is a vector quantity that's in the same direction as the velocity. The SI units for momentum are kg·m/s, and there's no commonly used derived unit that combines these units.

In a system of particles, in which each may have different amounts of momentum, the total momentum of the system is the (vector) sum of all the individual momenta. Unless acted upon by an outside force, the total momentum of the system remains constant. This is the principle of **conservation of momentum.**

The principle follows from Newton's second and third laws. A version of Newton's second law (equivalent to Equation 1.8) says that the net force on an object equals its rate of change of momentum. By Newton's third law, forces always come in pairs, with forces of equal strength directed oppositely.

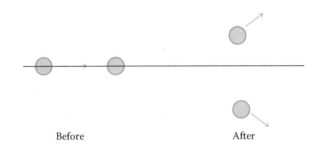

Before   After

Figure 1.3   Collision between two billiard balls, showing their motion before and after the collision.

Thus, the net force in any force pair is zero, so there can be no change in momentum for the system.

One useful application of the momentum conservation principle is in the analysis of simple collisions, like the one shown in Figure 1.3. Before the collision, one ball travels to the right, and the other one is at rest, so the total momentum of the system is just that of the moving ball, directed to the right. After the collision, the balls are moving in different directions. However, the total momentum of the two is equal to the momentum of the single moving ball before the collision. An example of where this kind of analysis might be useful is in the investigation of a collision between two moving cars. Using evidence of how the cars moved immediately after the collision, investigators can use the principle of momentum conservation to estimate how the cars were moving just before the collision. This can indicate which one was in the intersection illegally or traveling too fast.

How Do Rockets Work?

In a rocket, some kind of liquid or solid fuel is ignited. The rocket's engine channels the hot exhaust gases so that they're directed with a nozzle and ejected from the rocket. In doing so, the gases carry away momentum. Within the system of rocket plus gases, momentum is conserved, so the rocket accelerates in the opposite direction of the exhaust gases.

What Is Center of Mass?

Center of mass is a position within a system of particles given by the weighted average of all particle positions in a system. By weighted average, we mean that more massive particles count proportionally more in computing the average, so the center of mass tends to be closer to more massive particles.

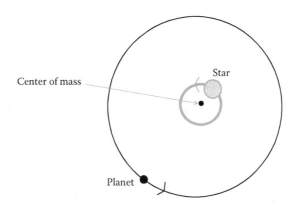

Figure 1.4   Motion of a planet and star around a fixed center of mass.

A corollary to the principle of momentum conservation is that the center of mass of a system can't accelerate unless the system is acted upon by outside forces. As an example, refer again to the billiard-ball collision in Figure 1.3. Before the collision, the center of mass is between the two balls and moves to the right. The collision can't change the center-of-mass motion, so the center of mass continues to move to the right after the collision, remaining between the two balls. Why then do they eventually stop rolling? Interactions with the table surface and side bumpers constitute outside forces, which can and do change the velocity of the center of mass of the two balls, eventually bringing it to rest.

Momentum transfer doesn't require a collision or direct contact. A single planet orbiting a star really orbits around their common center of mass, as does the star (Figure 1.4). The force of gravity that acts between them is a Newton's third law force pair. The force on each body changes its momentum continually. However, in the absence of outside forces, the center of mass remains in the same place.

WHAT IS SIMPLE HARMONIC MOTION?

An **oscillation** is any motion that proceeds back and forth over the same path, such as a swinging pendulum or the vibrating prongs of a tuning fork. The frequency of oscillation (symbol f) measures the number of times each second that a complete oscillation is made. **Simple harmonic motion** is an oscillation that follows a specific path, in which the motion varies sinusoidally in time. That is, the motion follows a sine or cosine function, as in

$$x = A \sin (2\pi f t) \tag{1.11}$$

where x is position along an axis, f is the oscillation frequency, and t is time. A is the amplitude of motion, measuring how far the oscillator moves in either direction from its central position. Figure 1.5 shows a graph of position versus time for a simple harmonic oscillator, with some of the key parameters shown on the graph. Notice that the motion is symmetric about the central position and the oscillator spends equal amounts of time on either side of center.

A model for a simple harmonic oscillator is a mass m attached to a spring, with the other end of the spring fixed. In this model the spring follows **Hooke's law,** which says that the spring exerts a force on the oscillating mass that is proportional to its displacement (x in Equation 1.11), and directed back toward the equilibrium position. Symbolically, the force from a Hooke's law spring is $F = -kx$, where the minus sign indicates that the spring always pulls or pushes the oscillator back toward equilibrium, with a strength proportional to the displacement, and k is a constant that measures the spring's stiffness. Mathematically, it can be shown that a perfect Hooke's law force results in the sinusoidal motion expressed in Equation (1.11).

No real springs follow Hooke's law perfectly, but many do so to a good enough approximation. These show up in a variety of physical systems. Atoms vibrating in a solid oscillate in a nearly simple harmonic fashion. Vibrating atoms in quartz crystals (solid SiO_2) are used to regulate many modern electronic clocks because they have such reliable vibration frequencies. Scientists study the behavior of molecules by measuring the energies of radiation they emit when they move from higher to lower energy states, corresponding to changes in vibration frequencies. On a macroscopic scale, vibrating strings in musical instruments follow a nearly simple harmonic pattern, and they generate sinusoidal sound waves (Chapter 3).

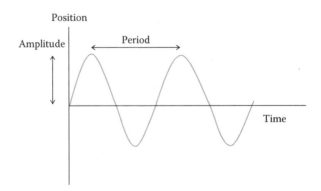

Figure 1.5   Position versus time for a simple harmonic oscillator.

WHAT ARE TORQUE AND ANGULAR MOMENTUM?

Think about how you turn a car's steering wheel. You apply a force, but not a push or pull. Rather, you turn the wheel by applying a force to its outer rim, nearly tangent to the surface. **Torque** (symbol τ, the Greek letter tau) is the physical quantity that measures the effectiveness of your action, and it's defined as the product of the force component tangent to the wheel multiplied by the radius. (We'll follow the physicist's practice of using Greek letters for most rotational quantities, to better distinguish them from the parameters of linear motion.) Because it's a force multiplied by a distance, torque has units N·m.

In linear motion, the effect of a net force is acceleration, with the relationship between the two given by Newton's second law (Equation 1.8). In rotational motion, the effect of a net torque is angular acceleration—a change in the rate of rotation. To make this a little more precise, we can define quantities analogous to the linear quantities of position, velocity, and acceleration. For rotational motion, **angular position** (symbol θ) is how far a rotating body has turned relative to some reference axis. **Angular velocity** (symbol ω) is the rate of rotation, which is the angle turned per unit time. **Angular acceleration** (symbol α) is the rate of change of angular velocity.

Thinking of Newton's second law ($F = ma$), you can see that torque is the rotational analog of force, and angular acceleration is the analog of linear acceleration. The rotational analog of mass is not simply the rotating system's mass because making something turn depends not only on its mass, but also on how that mass is distributed. The quantity that measures how the mass is distributed is called **rotational inertia** (symbol I), defined as

$$I = \sum_i m_i r_i^2 \tag{1.12}$$

This is a sum in which each bit of mass m_i has a corresponding rotation radius r_i relative to the rotation axis. With rotational inertia serving as the analog of mass for rotational motion, the rotational version of Newton's second law is

$$\tau_{net} = I\alpha \tag{1.13}$$

Think again of turning the steering wheel. The more torque you apply, the faster the wheel accelerates (in a rotational sense). The same goes for any rotational system.

Angular momentum is defined as $I\omega$, the product of rotational inertia and angular velocity. Note that it's also analogous to its linear-motion counterpart, linear momentum mv. The analogy continues in that net torque is equal to the rate of change of angular momentum, just as net force equals rate of change of linear momentum. Thus, a rotator with no net torque applied will maintain constant angular momentum. Planets orbit the sun with little change to their orbits from year to year because there are few objects that can provide enough torque to affect them much.

A familiar example of angular momentum conservation is the spinning motion of a figure skater. Being on skates reduces any applied torque to a minimum, so angular momentum can be maintained for a while. If a skater starts spinning with her arms extended and then pulls them in tight to her body, her spin rate (angular velocity) increases. That's because angular momentum is a product $I\omega$. Pulling in her arms reduces I, so ω must increase to keep the product constant. She can slow down by re-extending her arms, but can stop the spin only by digging her sharp skate blade into the ice, providing the needed torque to the system.

What Is a Gyroscope?

A gyroscope is a wheel that's freely turning but held in some fixed structure, like the one shown in Figure 1.6. The concepts of torque and angular momentum are useful in understanding how a gyroscope functions. A fast-spinning gyroscope can have a lot of angular momentum, so it will maintain a constant angular momentum and a constant spin axis unless external torque is supplied. That's why gyroscopes are essential components of navigation systems used on airplanes and spacecraft.

Figure 1.6   Photo of a gyroscope.

You may have played with a gyroscope similar to the one shown in Figure 1.6 as a toy. If you tilt it at some angle relative to vertical, gravity provides a torque that would make it fall over if it weren't spinning. Because it's spinning, however, the applied torque only changes the rotation axis, making the spin axis rotate in a cone-shaped pattern. This rotation of the spin axis is called **precession.** Our rotating Earth experiences a slow precession of its rotation axis caused by torque from the sun's gravity acting on Earth's slightly bulging equator. The effect is the well-known astronomical phenomenon of precession of the equinoxes, a slow drift of the apparent position of the sun relative to the background of stars.

WHAT IS NEWTON'S LAW OF UNIVERSAL GRAVITATION?

Newton's law of universal gravitation says that two small particles having masses m_1 and m_2 and separated by a distance r attract each other with a force of magnitude

$$F = G\frac{m_1 m_2}{r^2} \qquad (1.14)$$

where G is the universal gravitation constant, equal to about 6.67×10^{-11} N·m²/kg². Although the law technically applies only to small particles, the methods of calculus can be used to show that it also works for any spherically symmetric body—a good approximation for many stars and planets.

Why Is Newton's Gravitation Law Considered "Universal"?

You've probably heard that Newton has some connection to a falling apple. The idea is something like this: Newton saw the apple falling and at the same moment looked up and saw the moon, realizing that gravitational attraction to Earth is responsible for both motions—the falling apple and moon orbiting Earth. Whether or not the realization hit Newton in exactly this manner, the story illustrates the far-reaching effects of Earth's gravity. Similarly, the sun's gravitational force extends to planets as far as Neptune and to other bodies beyond. Immense galaxies—thousands of light years across—are held together by mutual gravitational attraction, and galaxies attract one another over even greater distances. Moreover, the force law and force constant G seem to be the same as far as we can see, although they're harder to test when we're looking at distant objects.

What Keeps Planets and Satellites in Orbit?

When a planet orbits a star (Figure 1.4) or a satellite orbits Earth, the orbital motion can continue for a very long time, unless the orbit is disrupted by an outside force. For an Earth satellite, higher orbits normally last longer than lower ones because, closer to Earth, drag forces from the atmosphere operate continuously on the satellite. This removes energy from the satellite, sending it to progressively lower orbits, and it eventually crashes to Earth. Falling satellite debris can be a serious hazard to life and property. Fortunately, it's statistically more likely that the debris will fall into the ocean or an unpopulated area.

What keeps a satellite moving around Earth? Newton's first law says that in the absence of external forces, an object will keep moving with constant velocity in a straight line. By Newton's first law alone, a satellite would soon move away from Earth. However, the attractive force of gravity pulls the satellite toward Earth. The combination of the satellite's inertia and acceleration toward Earth results in the closed orbital path. The same idea works for any orbital system, such as a planet orbiting the sun (Figure 1.7).

Launching a satellite from Earth's surface into orbit requires vertical motion to reach the desired altitude and horizontal motion with sufficient velocity to initiate the orbit. The amount of energy required depends on the satellite's mass and the eventual height of the orbit. Higher orbits require more energy.

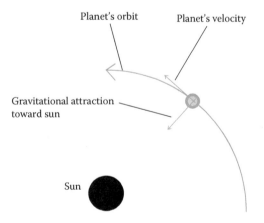

Figure 1.7   Orbital motion results from a combination of inertia and gravity.

What Orbits Do the Planets Follow?

Nearly a century before Newton, Johannes Kepler used some excellent naked-eye observations by his colleague, Tycho Brahe, to posit three laws of planetary motion. Kepler's laws are

1. Planetary orbits are ellipses, with the sun at one focus of the ellipse.
2. The orbital area swept out by a planet for a given time period is the same at any point in its orbit. (As a corollary, planets move faster when they are closer to the sun and slower when they are farther away.)
3. A planet's orbital period T and the semimajor axis a of its orbit are related, with T^2 proportional to a^3.

Kepler's laws are consistent with the observed orbits of all the planets, and they also work for the artificial satellites we have today. They work fairly well for the moon in its orbit around Earth. However, the moon's orbit is perturbed to a significant degree by the gravitational force of the sun. For the moon, Kepler's laws are still a good approximation, and discrepancies are explained by applying Newton's law of gravitation to take into account the sun–moon interaction. Similarly, planets' orbits around the sun are affected by other planets—especially by the heavyweight Jupiter—but Newton's law accounts for these attractions.

What Other Kinds of Orbital Paths Are Possible?

Kepler rightly identified an elliptical path as one likely shape for an orbit around a massive body. A circle is a noneccentric ellipse, so circular orbits are allowed too. A rigorous application of Newton's law of gravitation to the problem of orbits reveals that parabolic and hyperbolic paths are also allowed. As a group, circles, ellipses, parabolas, and hyperbolas constitute the four **conic sections** from geometry. There's a qualitative difference between these last two shapes and the first two: Circles and ellipses are closed, but parabolas and hyperbolas are open. Thus, only circular and elliptical paths make closed orbits that repeat. Many observed asteroids and comets follow hyperbolic paths around the sun, so we only see them pass through the solar system once. Many others have elliptical orbits, and we can see them repeatedly.

What Is a Geosynchronous Satellite?

Kepler's third law says that the higher a satellite's orbit is, the longer its orbital period will be. A low-Earth orbit, (i.e., at an altitude that's above just enough of the atmosphere to be sustainable) requires about 90 minutes. At a much higher altitude—a circular orbit with a radius equal to about 6.6 times Earth's radius—the period is 24 hours, making its orbit **geosynchronous.** In order to be **geostationary,** remaining over the same point on Earth's surface throughout its orbit, the geosynchronous orbit has to be over the equator. That makes the satellite useful for communication purposes because an Earth-based antenna can be fixed to point toward the satellite at all times.

Why Does the Acceleration due to Gravity Vary with Altitude and Latitude?

The acceleration a of an object with mass m in free fall near Earth is found to be setting the gravitational force (Equation 1.14) equal to ma (Newton's second law):

$$F = G\frac{m_E m}{r^2} = ma$$

where m_E is Earth's mass, and r is the object's distance from Earth's center. Thus, the free-fall acceleration is

$$a = \frac{Gm_E}{r^2} \qquad (1.15)$$

Note that this result is consistent with Galileo's observation about differently sized bodies falling at the same rate because the acceleration doesn't depend on m. Near Earth, r is approximately equal to Earth's radius, about 6.37×10^6 m. Using this number with Earth's mass, 5.97×10^{24} kg, gives an acceleration of about 9.81 m/s², a fairly typical value.

Moving from sea level to a higher altitude increases r (in Equation 1.15) and thus decreases the acceleration. Near Earth, the decrease is about 0.003 m/s² per kilometer of altitude gained. You can climb several kilometers up a mountain and not notice much difference, although you could detect the decreasing gravitational acceleration with sensitive instruments.

Earth's rotation gives it a slight bulge at the equator relative to the poles. This contributes to making gravitational acceleration lower at the equator. A second effect that contributes to lower acceleration at the equator is the **centrifugal force** that arises because a point fixed on the equator is in a rotating frame of reference. You may have sensed a centrifugal force while trying to hold on to the outer edge of a rapidly rotating playground wheel. You felt as if the wheel was trying to throw you off, although it's really your inertia that you sense, your tendency to travel in a straight line. Because the centrifugal force is not a real applied force but rather an artifact of a rotating frame of reference, it's called a **fictitious force.** A third effect at the equator, one that actually works against the other two, is a small increase in gravitational acceleration due to the nearby mass of the bulging planet. The net result of these effects is that gravitational acceleration is about 9.78 m/s² at the equator and 9.83 m/s² at the poles, with both measured at sea level.

What Is the Coriolis Force?

The **Coriolis force** is a fictitious force that arises in a rotating frame of reference. The effect on Earth is that objects moving horizontally relative to a fixed point on Earth curve to their right in the Northern Hemisphere and to their left in the Southern Hemisphere. The effect is normally small, but it can be noticeable when motion takes place over extended distances or time. One familiar result of the Coriolis force is the rotation of air masses around high and low pressure centers. Air generally moves from higher to lower pressure. In the Northern Hemisphere, the Coriolis force makes air rotate clockwise (viewed from above) around high-pressure zones and counterclockwise around lower pressure. In the Southern Hemisphere, the rotations are reversed.

The Coriolis force is responsible for the behavior of the **Foucault pendulum.** If a long pendulum is allowed to swing for many hours, the Coriolis force pushes the pendulum sideways with each swing. On each half-swing the force switches direction, causing the plane of the pendulum's swing to rotate slowly but steadily. A working Foucault pendulum provides direct evidence of Earth's rotation.

CAN CLASSICAL MECHANICS EXPLAIN EVERYTHING?

Newton's mechanics and law of gravitation greatly expanded the power of scientific thinking. Newton's work seemed to explain much of our universe, from the motions of planets to the mechanics of everyday life. In the years after Newton, the development of the concept of energy (not originally a Newtonian concept) and advances in mathematical methods (particularly those made by Lagrange and Hamilton) greatly increased the range of problems that could be addressed by classical mechanics. In the early nineteenth century Pierre Simon Laplace theorized that the determinism implied by Newtonian mechanics was so rigorous that the entire future of the universe could be predicted, provided the current positions and motions of particles could be measured with precision. A device that could accomplish such measurements and the computations to predict the future has come to be called a **Laplace's demon.**

We now know that a Laplace's demon cannot function and that classical mechanics alone is insufficient to explain the processes in nature. Even when classical mechanics is supplemented by classical electrodynamics (Chapter 2) to explain the behavior of charged particles and electromagnetic fields, an entirely new set of concepts and computations is needed to explain the subatomic world via quantum mechanics (Chapter 4). Quantum mechanics gives results that are sometimes inconsistent with classical mechanics, and in those cases quantum mechanics is correct. Further, quantum mechanics sets limits on the precision of measurement. It turns out that the initial measurements of position and velocity needed for a Laplace's demon to function can never be accomplished with sufficient precision, regardless of how those measurements are made.

Ultimately, the behavior of macroscopic bodies that seems to be well understood in classical mechanics is dependent on the behavior of the atoms and subatomic particles of which the larger bodies are made. At best, classical mechanics is only an approximation of the real universe. Because it's such a good approximation in many contexts, classical mechanics is still useful today, but we must appreciate its limitations.

FURTHER READINGS

Taylor, John R. 2005. *Classical Mechanics*. Mill Valley, CA: University Science Books.
Thornton, Stephen T., and Marion, Jerry 2003. *Classical Dynamics of Particles and Systems*. Belmont, CA: Brooks-Cole/Cengage.

Chapter 2

Electromagnetism and Electronics

In the twenty-first century, it's hard to imagine our lives without electronic devices. Most of us have become pretty dependent on our computers, cell phones, and various kinds of media players. Electronic devices are used by businesses not only for communication but also for design and manufacturing. Simpler electric-powered machines, from large motors to household appliances, have been around since long before the modern revolution in electronics. In this chapter we'll deal with some basic questions about electricity and magnetism, which are closely related to one another and have the same fundamental origin. Along the way we'll also address questions about some of the applications that touch your lives.

WHAT ARE ELECTRIC CHARGES, AND HOW DO THEY AFFECT EACH OTHER?

Electric charge is all around us—every atom in your body contains some electric charge! You don't notice it though because there are two kinds of charge, positive and negative, which cancel out in ordinary matter because they are found in equal amounts. The sources of all the charges you ordinarily experience are protons, with positive charge, and electrons, with negative charge. The charges on the proton and electron are said to be "equal and opposite": $+1.60 \times 10^{-19}$ C and -1.60×10^{-19} C, respectively. (The coulomb, symbol C, is the SI unit of charge.) Neutral atoms have equal numbers of protons and electrons. An atom with an imbalance of charge becomes an **ion,** with a negative ion resulting when an atom gains electrons and a positive ion if electrons are lost. Normally, only the electrons in an atom are mobile because the protons are locked inside the nucleus.

The simple rule for how the two kinds of charges interact with each other is that opposite charges attract and that like charges repel. You've seen this happen when you move charges around unintentionally. Comb your hair on a dry day, and you'll find that the rubbing action transfers electrons from your hair to the comb or vice versa. Either way, the originally uncharged hair and comb then have small net charges, opposite to one another. You know this because if you move the comb near your hair, you see the hair attracted to it.

Why Do I Get Shocked after Walking on Carpet?

You've probably had the experience of charging yourself by walking across carpet. To some degree the transfer of charge is simply because the two materials in contact with each other (shoes and carpet) have different affinities for gaining or losing electrons. The transfer that takes place upon contact is enhanced by the rubbing action. This helps to move some electrons either from the carpet to your shoes or vice versa, leaving you with a net charge. Then when you touch another person, attractive forces move some excess charge around to equalize it again, making an uncomfortable spark. It's remarkable that this works even though the other person may not carry a net charge. The organic molecules in your hand are uncharged, but they can be **polarized** by the presence of external charge, meaning that more of their charge is on one side than the other. If, for example, the external charge is negative, then the polarized molecule's positive end points that way. This in turn attracts the negative charge, making it jump from one finger to the other.

What Is Coulomb's Law?

Coulomb's law says that the force between two charges having magnitudes q_1 and q_2 separated by a distance r is

$$F = \frac{kq_1q_2}{r^2} \tag{2.1}$$

where k is a constant equal to about 8.99×10^9 N·m^2/C^2. With SI units, the charges are in coulombs (C) and the distance in meters (m), so the force comes out in newtons (N). The direction of the force on each charge is toward the other one if the charges have opposite signs and away from each other if the signs are the same. The force in Equation (2.1) is called the electric force, or electrostatic force ("static" because the force is present whether or not the charges are moving), or the Coulomb force.

How Strong Is the Electric Force Compared with Gravity?

Coulomb's law for the force between two charges looks remarkably similar to Newton's law of gravitation for the force between two masses. Both are inverse-square laws, meaning the force diminishes with the square of the distance. The product of the two charges in Coulomb's law resembles the product of the masses in Newton's law. However, there are a couple of important differences. One is that the electric force can be attractive or repulsive, depending on the signs of the charges, but the force of gravity is always attractive.

Another important difference is the strength of the forces of gravity and electricity. A good place to compare the two forces is in a hydrogen atom, where a proton and electron attract each other electrically and gravitationally. Using the two force laws with an average distance $r = 5.3 \times 10^{-11}$ m gives the two forces: 3.6×10^{-47} N for gravity and 8.2×10^{-8} N for electricity. That's an amazing difference of almost 40 orders of magnitude! This explains why electric forces dominate over gravity even when you have a fairly slight excess of charge, such as when you've run a comb through your hair. The big difference in the strength of the two forces also tells you that whenever gravity is the dominant force, the bodies involved are extremely neutral electrically. For example, Earth's orbit around the sun is accounted for by Newton's law of gravitation. That tells you that the net charge on Earth is pretty small.

What Is Quantization of Charge?

The electric charges of protons and electrons are responsible for all the electric charges you normally encounter. They both have the same magnitude of charge: $e = 1.60 \times 10^{-19}$ C, with the proton having charge $+e$ and the electron $-e$. We say that charge is **quantized** because all other collections of charge are multiples of the base amount $\pm e$. You can't break protons or electrons into smaller pieces. Physicists think of an electron as a fundamental particle that is not composed of any smaller ones. A proton is made of up of three quarks having charges $+2e/3$, $+2e/3$, and $-e/3$ (for a total of e), but the proton can't be split into quarks, for reasons we'll address in Chapter 8.

There are other exotic particles that are outside your everyday experience because they don't appear in ordinary matter and don't stick around long enough for you to notice. They include muons and positrons, which show up in cosmic rays and nuclear decays. Their charges are also quantized: $-e$ for the muon and $+e$ for the positron. Some particles have zero charge, such as the neutron that's found in the atomic nucleus (Chapter 7). There's a long list of other particles in nature, and they all have quantized charge—usually, $+e$, $-e$, or zero.

WHAT IS AN ELECTRIC FIELD?

Coulomb's law gives the force between two charges. In most situations involving charges, you have millions of them, so it's impractical to use Coulomb's law to find the net force on any one charge. In situations like this, physicists use the concept of a field to describe how the electric force acts on charges.

We'll introduce the idea of a field with the more familiar example of gravity. You know that if you drop an object from some height above Earth's surface, it experiences the force of gravity and begins to fall. You can think of the space above Earth having a **gravitational field** that will generate a force on any object you place there. In this case you'd define the field to have a value $g = F/m$, or the force per unit mass on the object. The field's direction is downward, toward Earth.

The **electric field** (symbol E) is defined in an analogous way, as the electric force per unit charge: $E = F/q$, measured in SI units newtons per coulomb (N/C). The direction of the electric field tells you direction of the force on a positively charged particle that's placed in the field. The force on a negative particle is directed opposite to the field. Figure 2.1 shows a good way to make a fairly uniform electric field: Distribute opposite charges (positive and negative) on a pair of parallel metal plates. The field (with its direction represented with arrows) between the plates points from positive to negative, consistent with the force that a positive charge would feel if you placed it in that space. The strength of the field depends on how much charge you put on the plates—more charge makes a stronger field, consistent with Coulomb's law.

What Happens in a Thunderstorm?

A bolt of lightning in a thunderstorm is an impressive electrical discharge. But how does the charge get there in the first place? Normally the charge buildup in a tall cloud is due to the combined presence of ice crystals and water droplets. Severe winds and turbulence cause frequent collisions between the water and ice, allowing some transfer of charge. After some time a strong electric field can build up within the cloud or between the cloud and ground. Once the electric field reaches a high enough value, called the **dielectric strength** of air (about 3×10^6 N/C), you get an electric discharge (**dielectric breakdown**), either from one cloud

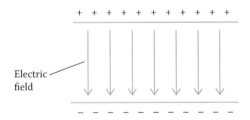

Figure 2.1   Electric field from a pair of parallel metal plates.

to another or a cloud to the ground. The amount of charge transferred—about 15 C for a typical bolt—is what makes the lightning display so impressive, and thunder comes from the resulting shock wave through the air. By the way, it's also dielectric breakdown that you feel when you shock yourself on a door handle. Thankfully, the amount of charge transferred is much smaller than in a lightning bolt!

WHAT IS ELECTRIC POTENTIAL?

Electric potential is a quantity related to electric forces and fields, and we'll use the familiar case of gravity again to explain. If you lift a brick from the ground, you give it some potential energy; the higher it goes, the more potential energy it has. Now let the brick go and it accelerates downward. Its potential energy decreases, but its kinetic energy increases. The total mechanical energy (sum of kinetic and potential) is conserved. An analogous thing happens if you release an electric charge (say a positive one) somewhere between the charged plates in Figure 2.1. Putting the charge in the field gives it some potential energy. Let the charge go and it accelerates toward the negative plate, converting some of the potential energy into kinetic energy. The moving charge's potential energy drops as it accelerates, just like the falling brick.

In this situation, **electric potential** is the potential energy divided by the charge. The SI unit for electric potential is the joule per coulomb, which is defined as a **volt** (symbol V). Dividing energy by the charge, you've defined a property of the space without the charge in it. Each point in the space has its own value of potential, which evidently drops as you get closer to the negative plate. The difference in potential (called, strangely enough, **potential difference**) between the two plates is a measure of how much energy is available to move charges around the space. You will often see potential difference—also measured in volts—in the analysis of electric circuits, and we sometimes refer to potential difference simply as "voltage."

WHAT IS A CAPACITOR?

A **capacitor** is a device used to store charge—normally equal amounts of positive and negative charge separated from one another. The charged parallel plates shown in Figure 2.1 make a good model to illustrate how a capacitor works, though in principle any two isolated conductors can be used to store the two kinds of charge. In that design, equal amounts of positive and negative charge are fed onto the two plates and held apart from one another. The opposite charges are attracted to one another but can't jump the gap between the plates, effectively storing the charge until it's needed to be pulled off and used. Capacitors are essential components in many electronic devices.

The ability of a capacitor to store charge is measured by its **capacitance** (symbol C), which is the ratio of the magnitude of the charge Q (positive and negative) stored to the potential difference V between the two charge-storing conductors. That is,

$$C = Q/V \qquad\qquad (2.2)$$

The unit for capacitance is the farad (F), with 1 F = 1 C/V. Capacitors in typical electric circuits are usually on the order of microfarads (μF = 10^{-6} F) to picofarads (pF = 10^{-12} F). The capacitance depends on the geometry of how the two charge-storing conductors are designed. Capacitance can be increased by placing a **dielectric** material between the conductors. A dielectric is an insulator that can be polarized. Its effect is to prevent an electric discharge across the gap while allowing more charge to be stored (for a given potential difference V), thus increasing the capacitance $C = Q/V$.

Although capacitors are found most often in electric circuits, some special capacitors are built for industrial or research purposes. At the National Ignition Facility in Livermore, California, high-powered lasers are used to initiate nuclear fusion (Chapter 7). This requires a large release of energy in a short time—in this case 2 MJ in about 1 ns. That's a fantastic rate of energy use: 2 MJ/1 ns = 10^{15} W, which is more than 100 times humanity's average energy consumption rate! Such a high rate is achieved by using large capacitor banks. The capacitors are charged slowly enough that the facility's energy use rate is reasonable; it's only the discharge that happens quickly.

How Does a Defibrillator Work?

A defibrillator is essentially a large capacitor that stores a lot of charge and then releases it quickly across a person's chest. That provides a jolt of electrical energy that can restart a stopped or irregular heart. A typical defibrillator is a 500-μF capacitor charged to a potential difference of about 1200 V. According to Equation (2.2), this releases (500 × 10^{-6} F)(1200 V) = 0.6 C of charge—a huge amount. The electrical energy associated with this release is over 300 J.

 WHAT ARE ELECTRIC CURRENT AND RESISTANCE?

Electric current is the flow of charge. Current (symbol I) is defined as the rate of charge flow (i.e., charge per unit time). The SI unit of current is the ampere (A), with 1 A = 1 C/s. In most situations, electric current is in the form of electrons moving through a circuit. Good conductors such as copper and other metals and alloys have a ready supply of electrons to carry current because the outermost electrons are very weakly bound to the atom. A potential difference applied to the conductor creates an electric field that forces the free electrons to move. Even though the charge carriers are (negative) electrons, current flow is normally given as a positive number (e.g., 1.5 A). However, the direction of this "positive current" in a circuit is opposite to the actual flow of electrons.

There's a vast difference in the ability of different materials to conduct electricity. We measure this ability using the quantity called **resistance** (symbol R), defined as the ratio of applied potential difference V to current I:

$$R = V/I \tag{2.3}$$

The SI unit of resistance is the ohm (symbol Ω, the uppercase Greek omega), with $1\ \Omega = 1$ V/A. From this definition, you can see that better conductors have lower resistance because they allow the most current for a given value of V. The reason for resistance is that electrons suffer inelastic collisions with the material as they propagate through it. The loss of energy shows up mostly as thermal energy, which is why you notice electronic devices getting warm when they're turned on. For a given amount of current, the rate of energy loss (P, for power dissipated) is

$$P = I^2 R \tag{2.4}$$

Broadly speaking, most materials fall into one of two categories: **conductors** (which conduct effectively) and **insulators** (which conduct poorly or not at all). There's not a distinct boundary between the two categories, but the difference in resistance between conductors and insulators can be many orders of magnitude. As we noted earlier, metals and metal alloys are generally good conductors because they have free electrons. Wood and plastics are common insulators.

What Is Ohm's Law?

Ohm's law says that the resistance of a given sample (as given by Equation 2.3) is constant and independent of the applied potential difference. In that case, $I = V/R$ is a linear function, as shown in Figure 2.2. A device that follows Ohm's law is said to be **ohmic**. In practice, no device is perfectly ohmic, though some materials follow approximately ohmic behavior for a wide range of currents. One reason conductors don't remain ohmic is that their temperature rises as more current passes through them. The resistance of a normal conductor increases with increasing temperature, which makes it non-ohmic. An example of this is the metal filament of an incandescent light, which gets quite hot under normal operation. The graph of current versus potential difference for such a filament is shown in Figure 2.3.

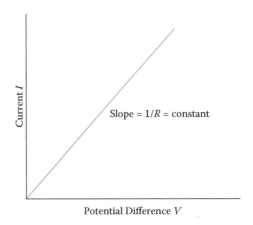

Figure 2.2 Graph of current versus potential difference for an ohmic material. The straight line indicates constant resistance.

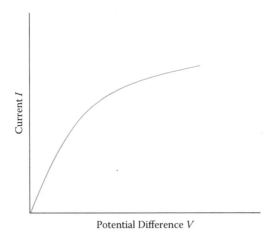

Figure 2.3 Graph of current versus potential difference for a tungsten light filament. With increasing current, the filament's temperature increases, raising its resistance. The changing resistance is revealed by the changing slope of the graph.

 ## WHAT ARE SEMICONDUCTORS?

A **semiconductor** material will conduct electricity, but it does so in a very different way than a normal conductor does. A normal conductor has free electrons that can move under the influence of an applied potential difference. In a semiconductor, electrons in an atom normally exist within a range of energies called the **valence band,** where they are attached to an atom and not free to conduct. In order to free themselves, they must jump across an **energy gap** (also called a **band gap**) to a higher energy range called the **conduction band.** The added energy is supplied through an applied potential difference, but no conduction is possible until a minimum potential difference is reached. As a result, a semiconductor's behavior is decidedly non-ohmic.

Another characteristic of semiconductors is that their resistance decreases with increasing temperature—the opposite behavior of normal conductors. In a semiconductor, higher temperatures provide thermal energy that enables more electrons to jump to the conduction band.

What Are Diodes and Transistors?

When an electron moves into the conduction band, it leaves a vacancy called a **hole,** which acts like a positive charge because the negative electron is absent. In an **intrinsic semiconductor,** there's a balance between negative electrons and positive holes. This balance can be changed by the process of **doping,** in which a small impurity is added to the semiconductor. If the impurity contains an excess of electrons, the semiconductor becomes **n-type** (for negative), and if the impurity is deficient in electrons, the semiconductor becomes **p-type** (for positive). A diode is a semiconductor device with n-type and p-type materials sandwiched together (shown schematically in Figure 2.4). The diode is an effective one-way device for current flow because there's a much greater tendency for electrons to flow from the n-type material to the p-type than vice versa. The current versus potential difference graph in Figure 2.5 shows this behavior.

Diodes are useful because they allow you to control the flow of current within a circuit. Even greater control comes from using a transistor, a three-terminal device that can be made of alternating semiconductor types: npn or pnp. Diodes and transistors are key components in advanced circuitry such as you find in modern electronic devices used for computing, communication, and graphic displays. In the early days of semiconductor design (the 1950s and 1960s), diodes and transistors were centimeter or millimeter sized. Since then they have been made smaller and smaller, so that now millions of them fit on a small circuit board that fits in your hand (Figure 2.6). These advances in semiconductor technology have made possible the recent revolution in consumer electronics.

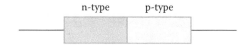

Figure 2.4   Schematic diagram of a diode.

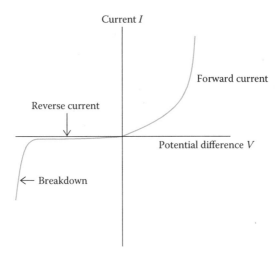

Figure 2.5   Current versus potential difference for a diode. Reverse current is severely restricted until breakdown occurs at a high reverse voltage.

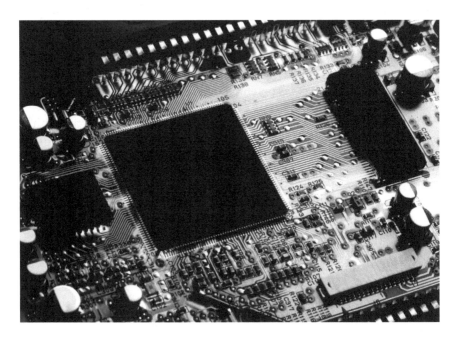

Figure 2.6   Photo of electronic components on a circuit board.

WHAT ARE SUPERCONDUCTORS?

The resistance of any material changes with temperature. Generally, a conductor's resistance drops appreciably as temperature falls, but good conductors such as copper and silver have some small resistance even at temperatures approaching absolute zero. Semiconductors have the opposite behavior. Their resistance falls as temperature increases because more electrons are thermally excited from the valence band to the conduction band.

Certain materials called **superconductors** do conduct electricity with zero resistance, but only at a sufficiently low temperature. Figure 2.7 shows what happens to the superconductor's resistance as its temperature changes. As the temperature falls, resistance drops gradually, as in a normal conductor, but at a **critical temperature** T_c (also called the transition temperature), the resistance drops suddenly to zero and remains there at any lower temperature. The critical temperature varies widely from one superconductor to another, ranging from just above absolute zero to over 100 K for some recently developed copper-oxide compounds. Currently, the highest T_c at atmospheric pressure is 135 K for a mercury-based copper oxide. There is evidence that some materials become superconducting at 164 K under high pressure. That's still more than −100°C, so any superconductor requires substantial refrigeration to operate.

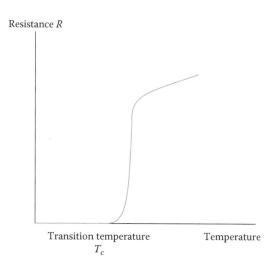

Figure 2.7   Resistance versus temperature for a superconductor.

What Are Some Applications of Superconductors?

Any resistance in a wire results in energy loss, so superconductors can be useful in applications that require substantial current flow. In any application, the gain from having no resistance loss must be balanced against the cost of refrigerating the superconducting material. That balance places some restrictions on the applicability of superconductors. Nevertheless, they are used extensively to carry high current in electromagnets that are used in research, industry, and medicine. If you've had a diagnostic test using **magnetic resonance imaging** (MRI), there's a good chance that a superconducting magnet was involved. The high quality of MRI images justifies the expense of cooling the superconductor. MRI uses the scientific technique of **nuclear magnetic resonance** (NMR), in which the magnet contributes to the absorption of radio-frequency radiation in nuclei. That absorption is highly dependent on the local environment, which is what enables MRI images to show fine detail from inside a person's body. Scientists use NMR to study the chemistry of new compounds, especially organic materials with hydrogen nuclei (protons), which are particularly susceptible to NMR.

Scientists use superconducting magnets for many purposes, most notably in large synchrotron particle accelerators such as the Large Hadron Collider at the CERN laboratory in Europe. There, magnets produce the high magnetic fields that are needed to bend high-speed protons around a large (27 km circumference) circular ring.

Another interesting application of superconducting magnets is in **maglev** (magnetic levitation), where large electromagnets suspend train cars above their rails. This reduces energy loss and makes for a smoother ride at high speeds. Several commercial maglev lines are open throughout the world, although not all of them use superconducting magnets.

Today a significant amount of electrical energy is lost due to resistance, both in its production and in transmission across long distances from power plants to consumers. Using superconductors in power plants and transmission lines would eliminate those losses. Another possibility not yet realized on a large scale is to use superconducting rings to store energy without loss, saving some energy produced during low-use periods for later, when energy needs are greater. Unfortunately, with presently available superconductor (and refrigeration) technology, none of these energy production, transmission, and storage systems are cost effective. Research may provide solutions in the years to come.

WHAT ARE MAGNETIC DIPOLES AND MAGNETIC FIELDS?

You've probably played with magnets of various shapes (Figure 2.8). Regardless of shape, a magnet generally has two identifiable poles, labeled north (N) and south (S), making it a **magnetic dipole.** At first glance this seems similar to

Figure 2.8   Two typical magnet shapes: bar and horseshoe magnets.

electric charges, which come in two kinds: positive and negative. Further, electric charges and magnetic poles follow the same rules of attraction to and repulsion from one another. For magnets, unlike poles (N and S) attract one another, and like poles (N and N or S and S) repel.

However, there's a big difference between electric charges and magnetic dipoles. You can't separate the N and S into separate pieces, the way you can with quantized positive and negative charges. If you try to cut your magnet in half, you won't make separate poles; you'll just have two smaller magnets, each having an N and S pole. Physicists describe this phenomenon by saying that there are no **magnetic monopoles.** The north and south poles you see on a magnet result from the combined effects of magnetic dipoles in individual atoms. The magnet you use to attract a paper clip harnesses together many atomic dipoles.

Analogous to the electric field established by electric charges, a magnet creates a **magnetic field,** illustrated in Figure 2.9. The magnetic field has a particular orientation, which tells you the direction in which a dipole placed in the field will align. The strength of attraction is governed by the magnitude of the magnetic field, measured in SI units of tesla (T). A 1-T magnet is considered strong, and a good electromagnet may generate 5 T. Like the electric field, the magnitude of the magnetic field drops off with increasing distance from the source, but for the magnetic dipole there's no simple relationship like Coulomb's law to connect the field strength to distance.

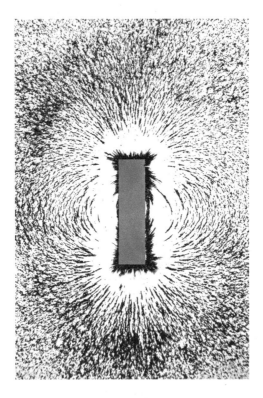

Figure 2.9   Small pieces of iron align with the magnet's dipole field.

What about Earth's Magnetic Field?

Navigators have known for many centuries about Earth's magnetic properties. A magnetic compass—essentially a magnetic dipole that's free to turn—aligns with the local magnetic field wherever you are on Earth's surface. The compass doesn't necessarily point directly north. Rather, it points toward Earth's north magnetic pole, which doesn't coincide with the geographic pole. The north magnetic pole is in reality the S pole of a dipole because the N pole on your compass is attracted in that direction: Not only is the magnetic pole not at the geographic pole, but it drifts in time. Its current location is in the Arctic islands of northern Canada. Thus, if you're in the eastern United States, your compass points west of geographic north, but in the western United States it points east of north. The source of Earth's magnetic field is believed to be electric current running through the outer core, which is composed of hot liquid iron. The current generates a magnetic field, which in turn helps keep the current going in a positive feedback cycle.

The magnitude of Earth's magnetic field varies somewhat across the surface of the globe, but typically it's about 5×10^{-5} T. That's easily strong enough to move a compass needle, but fortunately it's too weak to have an effect on most metal you carry around or on your electronic devices or credit cards. However, keep these things away from strong magnets. Your electronic gadgets and credit cards have magnetically coded memories that can be changed by a strong external magnetic field.

How Do Animals Use Earth's Magnetic Field?

Magnetic materials are used by a few animals. Some birds and bees contain small particles of magnetite (Fe_3O_4), which they use to help them navigate when visual clues aren't strong (e.g., at night). Some bacteria also contain small particles of magnetite, and they use it to move toward more desirable low-oxygen environments. There's recent evidence that some fish have magnetic particles in their noses, which they use as a navigation aid.

WHAT IS ELECTROMAGNETISM?

At first glance it may seem that electric circuits and magnets have nothing to do with one another and that electric and magnetic forces and fields are unrelated. It turns out, however, that electricity and magnetism are closely related on a fundamental level. Understanding how electricity and magnetism are joined in one phenomenon called **electromagnetism** not only led to a deeper understanding of physics but also made possible many of the applications in electronics and communication that you use every day.

What Is Magnetic Induction?

If a magnet sits next to a coil of wire, with both at rest, there's no interaction between the two. But if you put the magnet or coil in motion, an electric current begins to flow in the wire. This is an example of **magnetic induction.**

A more precise description of how this happens requires the concept of **magnetic flux,** which is proportional to the magnetic field and the surface area enclosed by the wire coil. According to **Faraday's law,** the current induced in the wire is proportional to the rate of change of magnetic flux. Anything that you can do to change the magnetic flux will induce a current. That's why it works to move either the magnet or coil. You can even leave the coil where it is and rotate it because that also changes the flux through it. But as soon as you stop all motion and bring both the magnet and wire to rest, the induced current drops to zero.

Magnetic induction is the basic tool used in most electric power generators. For example, in a hydroelectric dam, the mechanical energy from falling water is used to turn large turbines, which are connected to coils of wire that spin in a fixed magnetic field. The spinning induces current that's then sent over transmission lines for public use. Fossil-fuel power generators work the same way, except that the mechanical energy comes from steam generated by burning the fuel—usually coal or natural gas—that turns the turbine to induce electric current.

There are many other applications of magnetic induction. On the back of your credit or debit card is a magnetic strip, with information about your account encoded in the magnetic material. When you swipe your card in a reader, it passes the magnetic strip over a wire coil that's wound on an iron core. The magnetic material's motion induces current that carries your coded information.

Another important application is in a circuit component called an **inductor.** Think of an inductor as a coil of wire as part of a larger electric circuit. Any change of current in the circuit results in a changing magnetic flux in the inductor, which generates a current that affects the whole circuit. Properly placed inductors are thus useful in controlling current flow in electric circuits.

A **transformer** is an electrical device that relies on magnetic induction to change and regulate potential difference (voltage) in electric circuits. A model transformer consists of two coils with different numbers of windings wrapped around the same iron core. When the electric current in one coil changes, this generates a changing magnetic flux that's carried through the core, which in turn induces current in the other coil. The voltage in each coil is proportional to the number of loops of wire, so by using more or fewer loops, you can raise or lower the voltage. Designing the transformer properly allows you to generate the optimum voltage for running any electronic device. Transformers are an important component in electric transmission lines. The rate at which energy is lost due to electric resistance is proportional to the inverse square of the voltage. Thus, higher voltages (300 kV or more) are used for transmission lines from power stations to cities to reduce losses. Within cities, the transmission voltage is "stepped down" by transformers to 1–100 kV for safety, and then local transformers step it down to 120 V for consumer use.

How Does an Electromagnet Work?

An **electromagnet** uses the flow of electricity to generate a magnetic field. Any flow of current generates a magnetic field, but an electromagnet is designed to make a field of desired strength and shape in a defined space. Because an electromagnet uses the flow of electric charge and not magnetic materials like iron to generate the field, this might appear to be an entirely different phenomenon than a bar magnet. We'll stress again, however, that all electric and magnetic effects are closely related, so on a fundamental level there is only one phenomenon of electromagnetism, even if the effects are manifested differently.

A single loop of wire—even one carrying a large current—generates only a weak magnetic field. For that reason, a typical electromagnet consists of one or more coils of wire, each with hundreds or even thousands of loops of wire. A good electromagnet can generate a maximum magnetic field of 1–10 T. A single flat coil of wire produces a magnetic field like the one shown in Figure 2.10. The shape of the field resembles somewhat the dipole field of a bar magnet—an indication that the two different kinds of magnets are related on a fundamental level.

For large applications, superconducting electromagnets (discussed in an earlier section) are the best because they have no losses or heating due to resistance. However, they are also expensive to make and to keep cool. You can make an electromagnet with normal conductors, and you can run it at room temperature. Just remember that making a stronger magnetic field requires more electric current, so you will need a way to dissipate the heat that high currents generate in a normal conductor.

What Is a Solenoid?

A **solenoid** is a coil of wire wrapped in a cylindrical or helix shape (Figure 2.11). Extending the length of the coil has two effects on the magnetic field that's

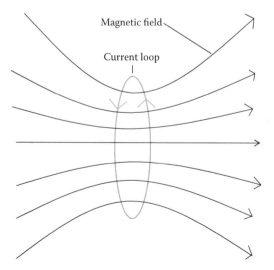

Magnetic field

Current loop

Figure 2.10 The magnetic field produced by a current loop.

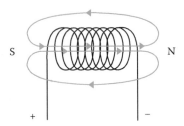

Figure 2.11 A solenoid is a coil of wire in the shape of a helix, producing a magnetic field that resembles that of a bar magnet.

produced when the solenoid carries a current. First, the magnetic field in the center is much more uniform than the field of a single loop or a flat coil (Figure 2.10). Second, the external magnetic field looks even more like the bar magnet's field, with the solenoid's length corresponding roughly to the bar's length.

A changing electric current in a solenoid produces a changing magnetic field that can be used to apply a force to a magnetic material that's partially in the field. This effectively turns some electrical energy into mechanical energy—the reverse effect of the electric generator. This design is used in a number of everyday applications. For example, when you start your car, a solenoid in your car closes an electrical contact that allows current to flow into the starter motor. Many doorbells use a solenoid that pushes a bar out to strike the bell whenever you push the button to close the switch. Solenoids are used to push metal bars to close electronic latches and locks. Another important household device that uses a solenoid is a **circuit breaker,** which activates whenever a maximum allowed current is exceeded. In this, the induced mechanical action opens a switch (instead of closing it) to prevent the flow of unsafe levels of current.

How Are Magnetic Materials like Electromagnets?

The charged particles (protons and electrons) that make up atoms all have an intrinsic angular momentum, called spin. A spinning charge acts like a current-carrying loop, so each electron and proton has an intrinsic dipole moment. Because this dipole moment is inversely proportional to mass, the much lighter electron is mainly responsible for the net dipole moment of atoms. Most materials don't exhibit obvious magnetic effects. For reasons that come from quantum mechanics (Chapter 4), electron dipole moments tend to cancel, either within individual atoms or among atoms throughout a bulk material.

However, in some materials the magnetic dipoles can add together to make a net magnetic moment for the material. That's what happens in your bar magnet, which is usually made of iron or an alloy of magnetic metals. Such materials are called **ferromagnetic,** or permanent magnets, although "permanent" is an exaggeration because the net magnetic moment runs down in time. Other materials, called **paramagnetic,** have a significant magnetic moment only when placed in a magnetic field. As a demonstration of paramagnetism, take a few paper clips. They aren't normally attracted to each other, but if you bring a bar magnet nearby, they gain a strong enough magnetic moment to be attracted to the magnet.

The root cause of magnetism in materials shows that there are not two kinds of magnetism—magnetic materials and electromagnets—but rather a single unified electromagnetism that is responsible for all magnetic effects.

WHAT ARE DIRECT CURRENT AND ALTERNATING CURRENT?

When electric current flows in one direction in a circuit, it's called **direct current** (**DC**). Current that reverses its direction periodically (at a regular rate) is **alternating current** (**AC**). Normally, AC current or voltage follows a sinusoidal pattern (Figure 2.12) with a constant frequency. Electric current delivered to your home is in the form of AC. In the United States the standard line is 120 V at a frequency of 60 Hz. In the United Kingdom and Europe, the standard is 230 V and 50 Hz. Japan uses 100 V; for historical reasons, the eastern part of Japan uses 50 Hz while the western part uses 60 Hz! The voltages

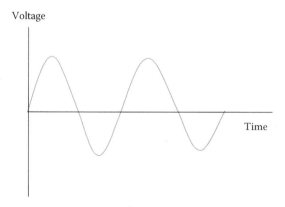

Figure 2.12   AC voltage as a function of time.

given here are root mean square (rms) values—effectively a time average. The peak voltage is a factor of $\sqrt{2}$ higher than the rms value, or about 170 V in the United States. These values can vary about 5% because the voltage drops as the net load on the system increases. However, the frequency—measuring the number of complete cycles of current or voltage per second—is held constant with high precision, in part so that electric clocks that use the AC signal will remain accurate.

Alternating current is easier to produce than direct current because any rotating-coil generator (discussed earlier) automatically generates AC. In the late nineteenth century a notable controversy arose regarding whether to use DC or AC for electric transmission. Thomas Edison, who counted the electric light among his many inventions, favored DC, and his name carried some weight. He was opposed by George Westinghouse and Nikola Tesla, who favored AC. AC is superior for transmission because it can be transformed to higher voltages to reduce resistive losses in transmission, and it eventually won out.

Tesla also worked out a scheme, which is still used, to transmit AC—not with the single voltage shown in Figure 2.12 but rather with a **three-phase system** in which three separate AC signals are delivered simultaneously, with each signal out of phase with the other two by one-third of a cycle. One advantage of three-phase AC is that the net voltage fluctuates less than in a single-phase line. This makes a smoother signal, which reduces vibration and wear in mechanical devices such as motors that run on electricity. Having three phases that follow a specific pattern also makes it easier to design motors that must turn in a single direction, like the ones in electric and hybrid cars.

Many machines and electronic devices require DC current. This means that the AC current you get from the wall outlet has to be changed to DC in a process called **rectification.** Rectifier circuits use the one-way properties of diodes to take an AC input and make it flow in one direction. Then, a carefully designed network of capacitors and resistors levels the time-varying current to form the constant DC output.

 ## HOW DO ELECTRIC MOTORS WORK?

An electric generator turns mechanical energy into electrical energy. An **electric motor** does the reverse, turning electrical energy into mechanical energy. The common DC motor has a source of DC current attached to a loop that's held in a magnetic field but free to turn. The force that makes the loop turn comes from the **Lorentz force**—the force experienced by any electric charge moving through a magnetic field. (This is yet another manifestation of the connection between electricity and magnetism!) The Lorentz force acts perpendicularly to both the magnetic field and the charge's direction of motion.

Hence, the forces on the two sides of the current-carrying loop are opposite, which produces a net torque on the loop to make it turn. A key component in the DC motor is the split-ring commutator, which makes the current reverse direction after each half-turn. That keeps the torque in the same direction as the loop turns.

WHAT IS AN ELECTROMAGNETIC WAVE?

Electromagnetic waves are the natural product of the connections between electric and magnetic fields. Imagine a single charge bouncing up and down in simple harmonic motion. The charge moving up and down constitutes electric current, which generates a magnetic field. Because the current is changing, the magnetic field does too and, as you've seen, that creates an electric field, which also changes. The changing electric and magnetic fields reinforce one another as they continue to move outward from the oscillating charge. It's those con-joined electric and magnetic fields that constitute the **electromagnetic wave** (or EM wave for short) illustrated in Figure 2.13.

One remarkable property of EM waves is that they all travel with the same speed in a vacuum: $c = 3.00 \times 10^8$ m/s, the speed of light. That's true regardless of the wavelength (peak-to-peak distance in the electric or magnetic field, as shown in Figure 2.13) or frequency. It's helpful to remember that for any travel-ing wave of fixed wavelength, the relationship between speed (c in this case), wavelength λ, and frequency f is

$$c = \lambda f \qquad\qquad (2.5)$$

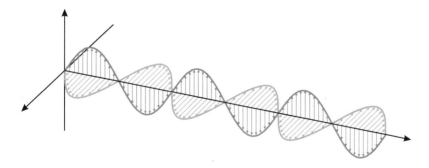

Figure 2.13   An electromagnetic wave has electric and magnetic fields at right angles to one another, with both at right angles to the wave's direction of motion, which is to the right in this diagram. The wave travels at speed c in a vacuum.

For an EM wave, c is constant, so the wavelength and frequency are inversely related. In the 1860s, physicist James Clerk Maxwell showed that the constant speed of EM waves is a consequence of the combined laws of electromagnetism. (Those laws are now called **Maxwell's equations,** appropriately.)

What Is the Electromagnetic Spectrum?

If electromagnetic waves travel at the speed of light, does that mean that light is really an electromagnetic wave? Yes, it is. However, what we know as visible light makes up only a narrow range of wavelengths, about 400–700 nm. The entire **electromagnetic spectrum** (or EM spectrum) is shown in Figure 2.14. Visible light is there, and notice that with wavelengths of 400–700 nm, the frequency of visible light (as given by Equation 2.5) is on the order of 10^{14} Hz. Beyond the boundaries of visible light lie the shorter **ultraviolet** waves and longer **infrared** waves.

At wavelengths longer than about 1 mm, you find all the radiation generically referred to as **radio** but including all the forms of communication we use for wireless communication—microwaves, AM and FM radio, television, radar, and so on. Historically, it was the discovery of radio waves by Heinrich Hertz in 1886 that confirmed Maxwell's analysis of electromagnetic radiation. In the early twentieth century, inventors made devices to communicate via radio signals over longer and longer distances, and the era of mass communication began.

At the other end of the spectrum from radio lie the waves with the shortest wavelengths and highest frequencies: x-rays and gamma rays. These extremely short waves are highly penetrating. You probably know that x-rays are used to diagnose medical conditions. Because they are absorbed or scattered differently

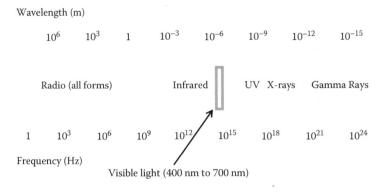

Figure 2.14   The electromagnetic spectrum.

by different kinds of body tissue, they are good for revealing internal structures. They can also be harmful, so you want to limit carefully your lifetime exposure to x-rays. With shorter wavelengths, gamma rays are even more penetrating and damaging. However, this is sometimes used as an advantage—for example, when gamma rays are used to kill cancer cells in radiation therapy.

Notice in Figure 2.14 that there are no fixed boundaries between the different parts of the electromagnetic spectrum. Even the boundaries of the visible spectrum are fuzzy because the ability to see at the ends of the spectrum varies from one person to another. Physicists consider it more important to distinguish between the types of EM radiation by how the radiation is produced rather than relying on wavelength or frequency. Most visible, ultraviolet, and infrared radiation comes from electronic transitions in atoms. Radio waves can come from smaller energy transitions in atoms, but they can also be made by oscillating charges in an antenna. X-rays are due to transitions between lower level electronic states, and gamma rays come from transitions in the nucleus. See Chapter 7 for more on x-rays and gamma rays.

How Does Wireless Communication Work?

As we described earlier, an oscillating charge pattern on a straight wire will generate an electromagnetic wave. The frequency of the oscillation governs the frequency of the EM wave. The radio part of the spectrum covers a wide range of frequencies, from a high of 300 GHz for a 1-mm wave to a low of several kilohertz for a 100-km wave. There are many schemes for transmitting and decoding data, but the general principle is that the message you want to send—sounds or data that will be translated into text or video—must be superimposed on the EM wave at the transmission stage and then extracted from the EM wave upon receipt. At the transmission end is an antenna to produce the charge oscillation, and on the receiving end a second antenna has its charges excited by the EM wave. Electronics in the receiver then remove the carrier EM wave to reproduce the message.

Over more than a century, wireless communication has evolved with improving technology. Different frequency ranges have various advantages, depending on the type of communication desired. For example, the "short wave" radio with frequencies of 3–20 MHz is useful for long-range communication because it reflects well off the atmosphere. Higher frequencies of 1–3 GHz are used for modern cell phone communication. This allows for greater clarity of signal and faster data transmission, but with limited range you have to be close to a cell tower to receive the wave. In practice, the allocation of frequencies is tightly controlled by national governments in the United States and elsewhere, to avoid a crowding of frequencies and interference between users.

How Does a Microwave Oven Work?

Microwaves with a frequency of 2.45 GHz (wavelength 12 cm) do a good job of exciting water molecules thermally because of the strong polar nature of the H_2O molecule. The oscillating electric field of the EM wave flips the molecules back and forth, and as they interact with one another the thermal energy spreads. Fortunately, the molecules in your container, such as a ceramic dish or cup, don't have those polar properties, so the microwaves pass through the container to your food or beverage. Reflections from the microwave oven's metal walls produce standing waves, which contribute to the energy transfer but can also leave cold spots in your food unless it's rotated.

It's remarkable that you can look inside the oven to see your food cooking without having the microwaves leak out and cook your flesh. Most microwave ovens have a wire mesh on the door, which is fine enough compared with the microwaves' wavelength that it effectively reflects them back into the oven.

FURTHER READINGS

Griffiths, David J. 1999. *Introduction to Electrodynamics*, 3rd ed. San Francisco, CA: Addison Wesley.

Jackson, J. D. 1998. *Classical Electrodynamics*, 3rd ed. New York: John Wiley.

Purcell, Edward M., and Morin, David J. 2013. *Electricity and Magnetism,* 3rd ed. Cambridge, UK: Cambridge University Press.

Chapter 3

Solids and Fluids

Life as we know it depends on the diverse properties of matter: solids and fluids. Just think about water in its various forms—liquid, solid (ice), and water vapor in our atmosphere, which eventually returns as rain or snow. And as a whole, the atmosphere contains a mixture of gases in an ideal combination to sustain life. Now think about the house you live in or the building you're in at this moment. Buildings are intricate mazes full of a variety of solid materials that serve different functions. Solids and fluids also carry waves, especially sound waves, as well as electric current. In this chapter we'll address some of the vital questions about how the properties of matter affect our lives.

 WHAT ARE THE STATES OF MATTER?

Common matter is found in three states: **solid, liquid,** and **gas.** Technically, the states are the forms that describe the three **phases** of matter, with each phase describing a range of conditions over which a material's behavior is uniform. The terms "state" and "phase" are often used interchangeably.

A solid has a definite shape that doesn't change unless extreme forces are applied. In a solid, the attractive forces between neighboring atoms or molecules hold them together to maintain the shape. The atoms are generally packed closely together, where interatomic forces (which are electrical in nature) are strongest. In a liquid, the atoms or molecules are still close enough that attractive forces keep the liquid together as a single substance of approximately constant volume—and hence uniform density. Liquids will flow to conform to the shape of their container. Gases also flow to fill their containers, but in a gas, intermolecular forces are so weak that the molecules are not bound to one another. Because gases have no definite shape or volume, their densities are highly variable based on conditions of pressure and temperature.

It's possible to distinguish between different states for the same solid form. For example, in a **crystalline** or polycrystalline material, the atoms in a solid form a regular pattern called a crystal lattice. In an **amorphous** solid such as glass, there is no such ordering of atoms. A solid electrical conductor can be in the normal conducting state or superconducting state (Chapter 2). Similarly, a liquid can act as a normal fluid or superfluid (discussed later in this chapter). Beyond solids, liquids, and gases, there are many other distinctions of "state," such as different magnetic states, liquid crystals, and Bose–Einstein condensates.

What Is Plasma?

An important fourth state of matter is **plasma,** an ionized gas in which electrons and nuclei exist together in a "soup" with the electrons not bound to the nuclei. You don't encounter plasma in your everyday life because creating plasma requires extreme temperatures. That excites the electrons thermally to the point where they aren't bound to nuclei. Plasma is found in the interior regions of stars, where temperatures range from thousands to millions of kelvins. Plasma is formed in many common electronic devices, including televisions and fluorescent lights.

How Are the States of Matter Related to Density, Pressure, and Temperature?

For a given material, density is normally greatest in the solid phase and least in the gas phase. This follows from the strength of the intermolecular forces corresponding to each phase. A material's density in the liquid phase is normally only slightly less than its density in the solid phase, and the gas's density is much lower but highly variable because it varies significantly with pressure and temperature. One notable exception is water, with a liquid density of 1000 kg/m³ and solid (ice) density of about 920 kg/m³ just below the freezing point.

A good way to understand phase transitions is on a thermodynamic **phase diagram.** Figure 3.1 shows the phase diagram for water. If you follow across the horizontal line at P = 1 atmosphere (atm; normal atmospheric pressure), you'll see water's familiar behavior. Starting at $T < 0°C$ and moving to the right, you cross the boundary between solid and liquid at 0°C; that is, ice melts. Continuing to higher temperatures, you see water boil—turning to its gas phase, steam—at 100°C. (The reverse processes for melting and boiling are freezing and condensation, respectively.) The phase diagram shows that the melting and boiling points are pressure dependent. If you live at high altitude, air pressure is less than 1 atm, so water freezes above 0°C and boils below 100°C. That's why cooking instructions call for longer cooking times at high altitudes.

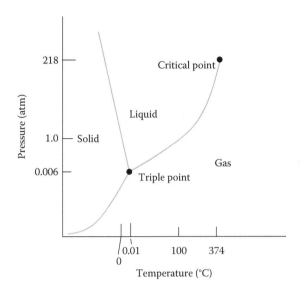

Figure 3.1   Phase diagram for water.

Two other notable points on the phase diagram are the triple point, where the three phases coexist, and the critical point, beyond which it's not possible to distinguish between the liquid and gas phases.

   Crossing the phase boundary directly from the solid to gas phase is called sublimation. You don't normally see this for water because, according to Figure 3.1, it can happen only at extremely low pressure. That's not the case for some other materials, notably carbon dioxide (CO_2), with the phase diagram shown in Figure 3.2. You've probably seen solid CO_2, commonly known as "dry ice," used for keeping things cold for shipping or storage. At atmospheric pressure, CO_2 sublimates at a temperature of $-78.5°C$. Such a low temperature is great for keeping things cold, and when it sublimates, the CO_2 gas released is harmless (in small enough quantities). Curiously, carbon dioxide has no liquid phase at all below 5.2 atm, so you aren't likely to encounter liquid CO_2!

What Makes Some Solids Stronger Than Others?

A solid's resistance to fracture and ability to bear weight are directly related to the bonding between atoms or molecules. For example, pure iron has a crystal structure that is moderately strong but can be made much stronger by the addition of carbon and other elements to form steel. Another example is pure carbon, which can exist in different forms having vastly different properties.

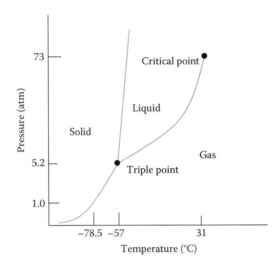

Figure 3.2   Phase diagram for CO_2.

When carbon forms bonds in a uniform, three-dimensional lattice, it makes diamond, one of the strongest solids known. Carbon can also bond in two-dimensional layers to make graphite, which is used in pencils because the layers easily slide off when you apply a slight pressure in writing. Other forms of pure carbon are the C_{60} **buckyball** and the **nanotube** shown in Figure 3.3, both of which were discovered recently and may have applications due to their strength and electrical conductivity.

WHAT IS A FLUID?

Because both liquids and gases are free to flow, they are grouped together as **fluids.** Although they are both fluids, liquids and gases differ greatly in their compressibility. Liquids are difficult to compress, but gases can generally be compressed easily. The ability to flow is responsible for many remarkable properties of fluids and leads to a great number of applications.

Why Does Fluid Pressure Depend on Depth?

You've no doubt felt the pressure change—maybe with your ears popping—when you go up or down in an elevator or airplane, and the sensation is even stronger when you dive a short distance under water. What you're experiencing is a change in **fluid pressure**, either in air or water.

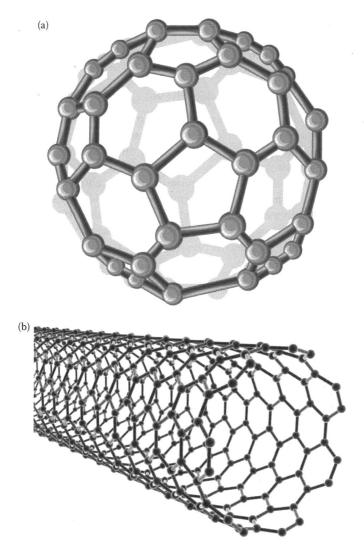

Figure 3.3   Two structures made of pure carbon: (a) C_{60} buckyball; (b) nanotube.

Your experience tells you that fluid pressure depends on depth. Why does this happen? As an example, think of the air in the atmosphere. We live at the bottom of an atmosphere that's many kilometers tall. (There's no way to specify an exact height because air density decreases gradually as you move away from Earth.) Pressure is force per unit area, measured in pascals (Pa), with 1 Pa = 1 N/m^2 in the SI system. Air pressure at Earth's surface is due to the weight of all the air above it, pressing down. As you rise in altitude, there's less air above you, so pressure decreases.

For a fluid with uniform density ρ, the relationship between changing pressure ΔP and changing height Δh is

$$\Delta P = -\rho g \Delta h \qquad (3.1)$$

where g is the acceleration due to gravity ($= 9.8$ m/s² near Earth). Note the negative sign in Equation (3.1), which is there because pressure increases as height decreases. As an example, think about pressure differences in the atmosphere. Typical air density near sea level is about 1.3 kg/m³. If you quickly rise 100 m in a tall building's elevator, the air pressure drops by about 1300 Pa, which is more than 1% of surface atmospheric pressure (about 100 kPa). That's enough for you to notice and maybe make your ears pop. Pressure differences in water are more significant because water's density is 1000 kg/m³. Diving just 2 m to the bottom of a pool increases the pressure by 20 kPa, or 20% of 1 atm!

How Does a Barometer Work?

A **barometer** is a device used to measure atmospheric pressure. The first barometers used a mercury-filled glass tube, as shown in Figure 3.4. The top of the tube is closed, and the bottom is open and placed in a pool of mercury. Air pressure pushes downward on the pool, but there's no air above the mercury in the tube, so the pressure there is zero. That pressure imbalance forces the mercury in the tube to stay at a height h above the pool. By Equation (3.1), there's a linear relationship between the pressure difference and height. Mercury is very dense (13,600 kg/m³), so the height of the column (corresponding to 1 atm of air pressure) is only 760 mm. Air pressure is often expressed in these non-SI units, millimeters of mercury (abbreviated as mm Hg).

In an **aneroid barometer,** which is in common use today, an evacuated cell (usually made of copper and beryllium) has a spring mechanism that allows the cell to expand or contract with changes in air pressure. The spring is usually attached to a dial, calibrated to read pressure directly. Common units for aneroid barometers are kPa, mm hg, or inches of mercury (in Hg), with 1 atm = 101 kPa = 760 mm Hg = 29.9 in Hg.

What Is Decompression Sickness?

Divers who go to significant depths experience increased pressure—about 1 atm for every 10 m of depth. At such high pressures, atmospheric gases dissolve gradually in the bloodstream and body tissues. If the diver returns too quickly to a lower pressure environment, those gases come out of the tissue and blood to form bubbles. This is **decompression sickness,** which causes

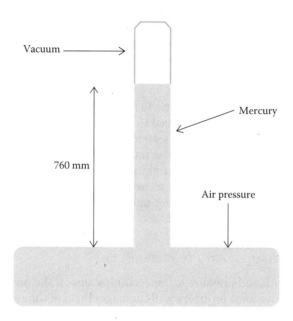

Figure 3.4   Schematic diagram of a mercury barometer.

sharp pains and can lead to tissue damage or death if gas bubbles block the flow of blood or other fluids. To keep this from happening, divers return to the surface on a timed schedule, holding for a time at specific depths on the way up. As an alternative, they can surface and enter a decompression chamber, a closed vessel in which the air pressure begins at a high level and is reduced gradually to 1 atm.

Decompression sickness is a hazard for those who are in any high-pressure environment. This includes mine workers—especially if the mineshafts have been pressurized to keep water out. It can also affect those who work in caissons—pressurized chambers in excavated areas below ground, such as tunnels and bridge foundations. Aircraft and spacecraft need to be pressurized to prevent sudden decompression that would cause similar physiological effects.

How Is Air Pressure Related to Weather?

On most weather reports, barometric pressure is given along with temperature, humidity, wind speed, and other data. That's because there's a strong correlation between atmospheric pressure and weather. High pressure usually brings clear or partly cloudy weather, with little chance of precipitation. Higher surface pressure is correlated to air that's predominantly sinking, which discourages cloud formation. Conversely, when the surface pressure is lower, water vapor can more easily rise to form clouds, so storms are associated with lower pressure. The worst storms—hurricanes and tornadoes—have very low pressure at their centers.

Air circulates in rotational patterns around high- and low-pressure centers, due to the Coriolis force (Chapter 1). In general, air flows roughly horizontally from higher to lower pressure regions. In the Northern Hemisphere, air (or any object) moving horizontally is pushed to the right, relative to its direction of motion, due to Earth's rotation. Thus, air leaving a high-pressure zone is always "turning right," setting up a clockwise rotation around the high-pressure center. Conversely, air turning to the right sets up a counterclockwise rotation around a low-pressure center. These rotations are clearly present in video images of clouds taken from orbiting satellites. In the Southern Hemisphere, the rotation directions are reversed.

What Does Your Blood Pressure Mean?

Measuring your blood pressure is important because, if the pressure gets too high, the excessive force on artery walls damages them in time and can lead to heart disease or stroke. The peak blood pressure, which occurs when the ventricles contract to force blood through the arteries, is called **systolic pressure.** When the ventricles relax, blood pressure returns to a lower value called **diastolic pressure.** The two numbers, reported as systolic/diastolic, are around 120/70 mm Hg for a healthy person.

To measure both systolic and diastolic pressure, an inflatable cuff is placed on the upper arm and inflated to pinch off blood flow. When the cuff pressure is gradually reduced, the pulse returns when the cuff pressure equals the systolic pressure. This is detected audibly, with a stethoscope, or electronically, as is done more commonly today. As the cuff pressure continues to drop, the audible pulse vanishes once the pressure reaches the diastolic value because then the arterial blood flow is smooth. In this way a single trial (letting the cuff pressure drift from high to low) gives both pressure readings.

What Causes Buoyancy?

Fluid pressure increases with increasing depth (Equation 3.1). That means that any object that's in the fluid feels more pressure pushing up from below than pushing down from above. This is **buoyancy,** and the net upward force due to the sum of fluid pressures is the **buoyant force.**

The amount of buoyant force is given by **Archimedes' principle,** which says that the buoyant force on an object in a fluid is equal to the weight of the fluid displaced. That's true whether the object is entirely immersed in the fluid or only partially, like ice floating in water. The principle is aptly named; Archimedes of Syracuse first announced it in the third century BCE.

Archimedes' principle explains why some things float. Think about an ice cube (density 920 kg/m³) in water (density 1000 kg/m³). If the ice is completely underwater, it experiences a buoyant force equal to the weight of the water it has displaced. Because of the density difference, the buoyant force is greater than the ice cube's weight. The net force on the ice is upward, and it rises to the surface. An ice cube floating on the surface remains at rest because with part of the ice sticking out above the surface, the buoyant force is less than when the ice is submerged. The floating ice is in equilibrium, with the buoyant force just enough to support the ice's weight. A careful analysis using Archimedes' principle shows that the fraction of the floating ice that's underwater is equal to the density ratio: 920/1000 = 0.92, or 92%. This ratio is just a little lower for an iceberg in seawater, which has a density of 1020–1030 kg/m³, due to the dissolved salts in it. This means that about 90% of the iceberg is underwater, so the "tip of the iceberg" really is only about 10%!

Archimedes' principle is what makes your helium-filled balloon rise and fly away when you let it go. In this case, helium's density, 0.18 kg/m³, is much less than air's 1.28 kg/m³. That's a huge difference, so the net buoyant force is strong, especially for larger balloons. That's why a helium-filled balloon or blimp can carry a heavy payload, including people. You want the net weight of your blimp—including payload—just to balance the buoyant force, so that the blimp will float at a constant altitude. For takeoff, air bags called ballonets (placed inside the blimp) are deflated to allow the helium to expand, which produces a small net upward force. For landing, the ballonets are inflated to compress the helium and create a downward force.

A hot-air balloon also employs Archimedes' principle. Warming the air lowers its density. Thus, the air inside the balloon weighs less than the cooler air it displaces. Of course, making the balloon rise requires a net upward force, so the buoyant force has to compensate not just for the air but also for the weight of the balloon fabric and any payload you want to carry.

What Is Bernoulli's Principle?

Bernoulli's principle is named for Daniel Bernoulli, who presented it in 1738, and it concerns the relationship between pressure and flow speed in fluids. In its simplest form, Bernoulli's principle states that increasing speed corresponds to decreasing pressure, and vice versa.

An important example is in aviation. Airplane wings are designed so that the incoming airstream is deflected, making a higher flow speed on the top of the wing than on the bottom. As a result, there's greater pressure on the wing's bottom, providing lift for the aircraft (Figure 3.5). Another example is in ball

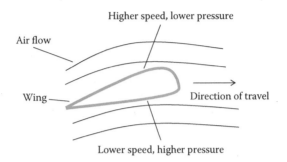

Figure 3.5   Lift on an airplane wing.

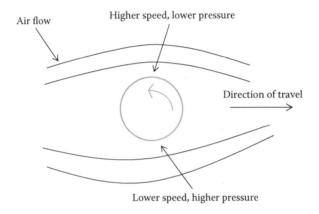

Figure 3.6   Sideways deflection on a spinning ball in flight.

games—soccer, baseball, tennis, golf, and many others—where a ball travel-ing through the air curves because of its spin. The spinning ball affects the air immediately around it, increasing the relative wind speed on the side of the ball rotating into the wind and reducing the wind speed on the opposite side. By Bernoulli's principle, there is greater pressure on the side moving into the wind, making the ball deflect sideways (Figure 3.6). A skillful player will con-trol the spin and make the ball curve in the desired direction.

What Is Surface Tension?

You've probably seen small objects resting on the surface of water, even objects such as paper clips that are denser than water and should sink. You're seeing an example of **surface tension,** which is the result of electrical forces

GOING DEEPER—BERNOULLI'S EQUATION

The simplified version of Bernoulli's principle can be made more quantitative with arguments based on conservation of energy. Sometimes it's important to extend this analysis to situations in which gravity is at work on the fluid. For example, think about the flow of water in a fire hose. The pressure and flow rate are related, but it also matters whether the fire is on the ground floor or high in the building because gravity will work to reduce the pressure the higher you go.

To begin the analysis, we'll apply the work–energy theorem (Chapter 1). The net work done on a fluid by external forces is equal to the change in its mechanical energy, kinetic ΔK plus potential ΔU:

$$W_{net} = \Delta K + \Delta U \tag{3.2}$$

From dimensional analysis, the work done on the fluid is the product of pressure (force per unit area) and volume, which gives the proper dimensions for work: force multiplied by distance. Bernoulli's equation relates fluid properties at two different places in the flow, which we'll label with subscripts 1 and 2. Thus, the net work in Equation (3.2) is

$$W_{net} = P_1 V - P_2 V$$

for an incompressible fluid with volume V.

The change in kinetic energy in Equation (3.2) is

$$\Delta K = \tfrac{1}{2}mv_2^2 - \tfrac{1}{2}mv_1^2 = \tfrac{1}{2}\rho V v_2^2 - \tfrac{1}{2}\rho V v_1^2$$

where we've used the fact that mass = density × volume.

Finally, the change in potential energy in going from height h_1 to h_2 is

$$\Delta U = mgh_2 - mgh_1 = \rho V g h_2 - \rho V g h_1$$

Putting the work, kinetic energy, and potential energy results into Equation (3.2) and rearranging gives

$$P_1 + \tfrac{1}{2}\rho v_1^2 + \rho g h_1 = P_2 + \tfrac{1}{2}\rho v_2^2 + \rho g h_2 \tag{3.3}$$

Note that the volume V was present in every term and canceled out. The result, Equation (3.3), is called **Bernoulli's equation** and is a general result for an incompressible fluid. Not surprisingly, Bernoulli's equation tells us that the water pressure in your fire hose drops as you go up in height. In fact, you can see that the static pressure of a fluid as a function of depth (given in Equation 3.1) is just a special case of Bernoulli's equation when the fluid isn't moving. Our earlier analysis of the airplane wing—where higher pressure is associated with slower flow rate—is the special case where the height difference is negligible.

between the polar water molecules. Under the surface, the water molecules are in equilibrium because they are attracted to their neighbors in all directions. Surface molecules are also in equilibrium, but with no neighbors on one side, forces parallel to the surface dominate, making the surface slightly elastic. This enables the surface to support small objects. Even some small insects can walk across the surface! Surface tension tends to minimize the water's surface area, which is why water tends to form spherical drops. Falling raindrops are actually fairly spherical, not the "teardrop" shape you often see depicted.

Capillary action is related to surface tension. You may have noticed that liquid in a graduated cylinder tends to curve upward at the edges. This happens when adhesive forces between the cylinder and liquid overcome the liquid's surface tension. In a cylinder filled with mercury (as in a barometer), the mercury and glass are not attracted at all, so surface tension makes the mercury's surface curve the other way—upward in the middle. Capillary action becomes more pronounced in thinner tubes (hence the name *capillary*) and, in a thin enough tube, a liquid can creep a considerable distance.

What Is a Superfluid?

Fluids in contact with matter experience **viscosity**—resistance to flow. At extremely low temperatures, materials called **superfluids** lose all resistance to flow; that is, they have zero viscosity. The most common superfluid was also the first one discovered: liquid helium, which behaves like a normal fluid from its boiling point of 4.2 down to 2.2 K. At 2.2 K and lower, it's a superfluid. (These temperatures assume the helium is at a pressure of 1 atm; the transition temperatures change when the pressure changes.) Physicists describe the transition from a normal fluid to a superfluid as a phase transition, analogous to the transition that occurs when a substance melts or boils. In this case, the phase transition is manifested by a sharp spike in the helium's heat capacity near the transition and sharp drop in entropy from the normal to superfluid state. Quantum physics has provided a theoretical description of a superfluid as an example of a **Bose–Einstein condensate,** in which a large number of particles condense into a single quantum state.

Superfluid helium has some remarkable properties. With zero viscosity, it can easily pass through microscopic pores and capillaries. It even forms a "creeping film" that can flow upward on the vertical walls of its container due to surface tension.

WHAT ARE TRANSVERSE AND LONGITUDINAL WAVES?

Everyone is familiar with waves traveling in water. In general, a wave is some form of disturbance traveling in a medium. In water waves, alternating bands of higher regions (wave crests) and lower regions (wave troughs) travel together in groups. A water wave is an example of a **transverse wave,** which means that the disturbance is perpendicular to the direction of travel. In this case it's the vertical water displacement that's perpendicular to the generally horizontal direction of the wave's travel.

Water provides a good model for defining some important wave properties, illustrated in Figure 3.7. The **wavelength** (symbol λ) is the distance between successive wave crests (or troughs). The **frequency** f is the number of whole wavelengths passing a fixed point per unit time. With dimensions of 1/time, the units for frequency are cycles per second (s^{-1}), also called a hertz (Hz). The **amplitude** A refers to the height of the disturbance. More precisely, A is usually defined to be half of the vertical distance from crest to trough (Figure 3.7). From dimensional analysis, the wave's speed v is related to its frequency and wavelength by

$$v = \lambda f \qquad (3.4)$$

Equation (3.4) is an important rule for waves. Wave speed is often fixed by the nature of the wave-carrying medium. For a fixed speed, Equation (3.4) tells you that a longer wave has a lower frequency and a shorter wave a higher frequency. That makes intuitive sense, based on how frequency is defined.

Sound traveling through air provides a good model for a **longitudinal wave,** in which the disturbance and wave motion are in the same direction. A sound

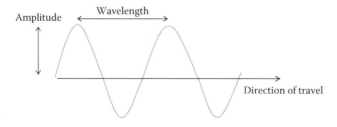

Figure 3.7   Wave properties.

wave contains regular variations in air density, with higher density compressions alternating with lower density rarefactions. (Sound waves can also travel though liquids and solids, where they are longitudinal waves that correspond to variations in pressure but can also include a transverse component.) The wavelength of a longitudinal wave is the distance between successive compressions or successive rarefactions, and the frequency is defined in the same way as for a transverse wave: the number of whole wavelengths passing a fixed point per unit time. Defining the amplitude of a longitudinal wave depends on the type of wave and medium.

Transverse and longitudinal waves look very different, but they share the common feature of transporting energy from one place to another without transporting matter. Light and other forms of electromagnetic radiation (Chapter 2) transport energy using via electromagnetic waves, which are transverse.

What Is Wave Interference?

Interference occurs when two or more waves meet and overlap in a region of space. There are many possible effects, depending on the kinds of waves that interfere. For most waves, the mathematical description of two-wave interference is simply the sum of the two individual waves, and this sum is called the **superposition** of waves. Interference of waves may result in constructive or destructive interference. In **constructive interference,** the wave maxima add together, resulting in a wave having greater amplitude than either of the individual waves. In **destructive interference,** a maximum from one wave overlaps with a minimum from another, resulting in a partial or complete cancellation of the amplitude.

When two light waves of the same wavelength interfere, the result may be a pattern containing regions of constructive interference alternating with regions of destructive interference (Chapter 5). A common example of interference is the phenomenon of **beats** in sound waves. This occurs when two sound waves having slightly different frequencies interfere—for example, two musical instruments that are slightly out of tune. For a fixed listener, the sum of the two waves varies in time between constructive and destructive interference, and you hear the corresponding increase and decrease in volume, varying regularly in time. A mathematical analysis shows that the frequency of the volume variation is equal to the difference between the two source frequencies.

What Is a Standing Wave?

A **standing wave** results from interference between two waves of the same frequency that travel in opposite directions. A good way to make a simple standing wave is to take a string or coiled spring that's fixed at one end and shake the other end with a constant frequency and amplitude. In this case, interference occurs between the generated wave and its reflection from the fixed end. Different standing wave patterns are possible, depending on the generating frequency. What distinguishes a standing wave is the existence of **nodes,** which are still points created by destructive interference, and **antinodes,** where the waves can add constructively to maximum amplitude (Figure 3.8). The number of nodes and antinodes increases as the generating frequency is increased.

Standing waves can form in a microwave oven, which uses microwaves (electromagnetic radiation) with a frequency of 2.45-GHz and a wavelength of 12.2 cm. (Note that Equation 3.4 gives the microwave's speed—about 3.0×10^8 m/s, which is the speed of light.) The 2.45-GHz frequency is readily absorbed by water molecules in the liquid state, raising the temperature of your food or drink. Microwave ovens have metal walls, which reflect unabsorbed microwaves back into the oven. There they can interfere with one another and create standing waves. Any part of the food placed at a node will warm at a much slower rate. Rotating the food allows each part of it to pass through nodes and antinodes, resulting in more uniform warming.

What Is a Tsunami (Tidal Wave)?

A **tsunami** (from the Japanese for "harbor wave") is really a series of waves generated by a significant disturbance of the ocean—usually from an earthquake, though they can also be caused by volcanic eruptions or landslides. Tsunamis

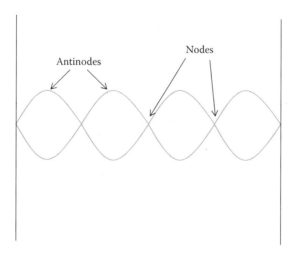

Figure 3.8 Illustration of a standing wave pattern. This standing wave appears on a horizontal string held between two fixed boundaries.

are unlike waves that regularly occur on the ocean in that they have much longer wavelengths and often much larger amplitudes. The slowly rising water in the long wave can resemble an incoming tide, giving rise to their description as "tidal waves." Because of height and length, a tsunami contains an enormous volume of water and can penetrate far inland in a low-lying coastal area, causing significant loss of life and damage to physical structures. Two significant tsunamis in this century—both caused by earthquakes—were the one that struck the Indian Ocean in 2004 and the one that struck the western Pacific Ocean off Japan in 2011. The 2004 tsunami was caused by an earthquake of magnitude 9.2, one of the largest in recent years. The 2011 tsunami was noteworthy because in addition to the damage and loss of life it caused on the Japanese coast, it triggered a serious accident at a nuclear power facility in Fukushima.

HOW IS SOUND GENERATED?

A sound wave consists of variations in air density. You're familiar with many ways of making sound. Vibrations in matter are particularly good for generating sound. For example, vibrations in your vocal cords generate sound waves in air when you speak. Vibrating strings are the sources of sound in many musical instruments. Drop an object on the ground, and the collision deforms the molecular structures in the object and the floor, making them vibrate and sending sound waves through the air to your ear. Sound can travel in any medium that can carry a density wave, including solids and liquids.

What Is the Speed of Sound?

Sound propagates with a speed that depends on the medium and conditions. In air the speed of sound is about 331 m/s at 0°C and 1 atm pressure. With changing atmospheric conditions, the speed of sound varies with the inverse square root of density. Thus, the speed of sound increases as air becomes less dense. This corresponds to a fairly linear increase in the speed of sound with increasing temperature, rising from 331 m/s at 0°C to 343 m/s at 20°C. Increasing humidity also makes sound travel faster because humid air is less dense than dry air. In other gases, density strongly affects the speed of sound. At 0°C and 1 atm, the speed is about 260 m/s in CO_2, which is denser than air, but about 960 m/s in helium, which is much less dense.

Sound generally travels faster in liquids and solids than in gases. It's about 1400 m/s in water and even higher in most solids—for example, 3000–4000 m/s in wood, 3900 m/s in copper, and 6400 m/s in aluminum.

What Is the Decibel Scale?

Sound waves carry energy. A direct measurement of how much sound energy you're getting is the **sound intensity** I, defined as the power per unit area at a particular receiver location. (Remember that power is energy per unit time.) As you move away from the sound's source, the area increases as the square of the distance. Therefore, intensity falls with the inverse square of the distance. Move twice as far away, and the intensity drops by a factor of four.

Intensity isn't the best indication of how loud you perceive a sound to be because human hearing doesn't react linearly with intensity. Instead, we perceive sound getting louder by a similar amount each time the intensity jumps by a constant factor. The commonly used **sound intensity level** β is defined as $\beta = \log(I/I_0)$, where $I_0 = 10^{-12}$ W/m² is a threshold hearing level for a person with excellent hearing. The unit for β is the bel (B), named in honor of sound researcher (and telephone inventor) Alexander Graham Bell. It's more convenient to use units of 0.1 bel or **decibel** (dB), which makes the sound intensity level

$$\beta \text{ (in decibels)} = 10 \log(I/I_0) \tag{3.5}$$

For example, typical intensity that you hear in normal conversation is about 10^{-7} W/m², corresponding to 50 dB. Ten times the intensity (10^{-6} W/m²) has a sound intensity level of 60 dB. Every 10-fold increase in intensity adds another 10 dB. High decibel levels can be painful and damaging to your hearing (Table 3.1).

How Does Your Hearing Work?

Sound is created when some mechanical motion generates sound waves in the air. In your ear, that process is reversed when sound waves are converted back into mechanical motion that your brain can interpret.

TABLE 3.1   DESCRIPTION OF SOME BENCHMARK SOUND INTENSITY LEVELS

Sound Intensity Level (dB)	Description of Sound
0	Hearing threshold
20	Whisper
50	Typical conversation
80	Busy city street
100	Near loudspeaker at concert
120	Near jet aircraft at takeoff
160	Eardrum rupture

First, sound waves are collected and focused by the curved, fleshy part of your ear that lies outside your head. Sound waves pass through a narrow ear canal and strike the eardrum, which vibrates at a frequency and amplitude that depend on the frequency and intensity of the sound. The vibrating eardrum moves against a series of three bones, which serve to amplify the incoming signal into the fluid of the inner ear. That fluid vibrates in chambers that contain small hairs, which transmit the signal to the auditory nerve and the brain.

Humans with good hearing can hear sound intensity levels above 0 dB, but the ability to hear also depends on the sound frequency, especially at lower intensity levels. The range of human hearing can cover frequencies from 20 Hz to 20 kHz, although this varies from person to person. The range generally shrinks as part of the aging process, with higher frequencies affected the most. Some animals—for example, marine mammals and bats—use **ultrasonic** frequencies (above 20 kHz) for communication and range-finding. We can generate and use ultrasonic frequencies, even if we can't hear them. Ultrasound imaging is now a common medical tool. Remember that the higher a wave's frequency is, the shorter its wavelength will be. The 10-MHz waves typically used in medical imaging have wavelengths of less than 1 mm in the body's fluids, providing good detail on the images that are produced by computer analysis of the reflected ultrasound waves (Figure 3.9).

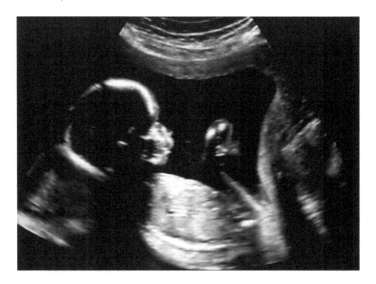

Figure 3.9   Ultrasound image used in medicine.

HOW ARE MUSIC AND MUSICAL HARMONIES MADE?

A vibrating string is a good start in making a musical instrument because its vibration frequencies can be changed (tuned) by changing its density, length, and tension. The ancient Greeks made stringed instruments, and Greeks in the Pythagorean school noted how pleasing, harmonious sounds resulted when they played simultaneously two similar strings with lengths in whole-number ratios, such as 2:1 and 3:2. In doing so they produced sound frequencies in those ratios, and it's the frequencies that make different musical notes. What we call low and high "pitch" is really low and high frequency of the sound waves. We sense harmony when the frequencies are in those whole-number ratios—2:1, 3:2, 5:4, and so on.

The frequency ratio 2:1 is defined as an interval called an **octave,** and this serves as the basis for the musical note scale used by musicians. The 440-Hz note is defined as an "A" on the musical scale. One octave higher is 880-Hz and is also called A. In-between there are 12 steps, in increasing order: B-flat, B, C, and so on, with the 12th step the next A. Each of the 12 steps increases the note's frequency by the same factor, $2^{1/12}$ (about 1.0595), so that after 12 steps, the octave is reached because $(2^{1/12})^{12} = 2$. Playing the right notes together makes harmonies. For example, playing the C with an E, which is four steps higher, makes a ratio $(2^{1/12})^4 = 2^{1/3} \approx 1.26$—close enough to 5:4 in whole numbers.

When playing a particular note, musical instruments don't make a single frequency. A violin playing a 440-Hz "A" produces not only that frequency but also many others, caused by secondary vibrations in the string and the wooden box below, which is carefully crafted to generate a rich combination of frequencies. The same "A" note on a clarinet sounds very different and is easily distinguished from the violin's note by your ear. The clarinet's sound pattern is further enriched by the fact that it originates from a thin vibrating reed (activated by the musician blowing across it) and includes resonances in a long wooden tune. Brass instruments involve blowing into a cupped mouthpiece, with resonances forming in long, curved metal tubes. Within any musical family, the size of the instrument generally affects wavelength, which is inversely related to frequency. For example, the large string bass makes longer waves and lower frequencies than the violin.

Percussion instruments are a special category. Drums have a tightly stretched sheet that, when struck, supports a sometimes complex two-dimensional wave pattern. That wave in turn makes the sound waves you hear. Larger drums tend to make lower frequencies. Other percussion instruments without drumheads involve various materials and shapes that make sound when struck—think of triangles, xylophones, and cymbals.

WHAT IS THE DOPPLER EFFECT?

You may have noticed that when a train or car is moving toward you and then passes and starts moving away, there's a noticeable change in pitch (frequency) in the sound you hear from the vehicle's engine or horn. That's the **Doppler effect**—a change in the frequency of sound received due to the relative motion of source and receiver.

Figure 3.10 shows how the Doppler effect works. When there's no relative motion between the source and receiver, the source frequency f is received at the same frequency. When the source is moving, wave crests are bunched in the forward direction and spread out behind, resulting in shorter or longer wavelengths, respectively. The shorter wavelength corresponds to a shift to higher frequency, and the longer wavelength means a lower frequency on that side.

Note that although Figure 3.10 shows the source moving and receivers stationary, it works just the same if the source is stationary and the receiver is moving toward or away from it. It also works if both the source and receiver are moving. In each case, the Doppler effect is due to the *relative* velocity between source and receiver.

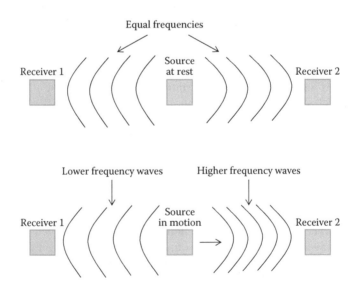

Figure 3.10   The Doppler effect results in higher or lower frequencies received due to the relative motion of source and observer.

GOING DEEPER—THE DOPPLER EFFECT

Analyzing the wave pattern in Figure 3.10 allows you to relate the frequency f' of the received sound, relative to the source frequency f, as a function of the source velocity v_s. The result is

$$f' = \frac{f}{1 \pm v_s/v} \tag{3.6}$$

where v is the speed of sound. The $\pm$ symbol refers to whether the source's relative motion is toward (−) or away from (+) the receiver. The − sign results in a higher frequency and the + sign in a lower frequency, consistent with Figure 3.10.

Equation (3.6) shows that it's the ratio of the source speed to the speed of sound that determines the amount of frequency shift. For example, suppose a car is traveling at 100 km/h (27.8 m/s) and sounds its 400-Hz horn. If the car is moving toward you, Equation (3.6) shows that you'll hear the sound at a frequency of 435 Hz (assuming a typical sound speed of 343 m/s), a significant frequency shift. If the same car is moving away from you, you'll hear the sound at 370 Hz. If the same horn is on a bullet train traveling at 250 km/h (69.4 m/s), the shifted frequencies are 501 and 333 Hz. This shows the effect of increasing speed.

WHAT MAKES A SONIC BOOM?

Referring again to Figure 3.10, think about what happens when the source's speed is very close to the speed of sound. Then the waves bunch up tightly ahead of the source, creating a **sound barrier** with increased drag force. If that barrier can be overcome and the source's speed exceeds the speed of sound, the bunched sound waves are released in a **shock wave,** which you hear as a **sonic boom.**

Similar shock waves form for other kinds of waves, whenever the wave generator is moving faster than the wave itself. One example of this is bow waves from a fast boat. Another example from physics is when fast-moving particles exceed the speed of light in the medium through which they're passing. (This doesn't break the law of relativity, which says that a particle can't exceed the speed of light in a vacuum.) The result—analogous to the shock wave—is an intense beam of electromagnetic radiation called **Cerenkov radiation.**

FURTHER READINGS

Berg, Richard E., and Stork, David G. 2004. *The Physics of Sound*, 3rd ed. San Francisco, CA: Addison Wesley.

Crawford, Frank S. 1968. *Waves* (Berkeley Physics Course). New York: McGraw–Hill.

Rex, Andrew, and Wolfson, Richard 2010. *Essential College Physics*, vol. 1. Boston, MA: Addison Wesley.

Chapter 4

Quantum Mechanics

Quantum mechanics (or simply quantum physics) is the branch of physics that applies to particles and systems of particles, such as atoms and molecules. In that realm the laws of classical mechanics and classical electromagnetism break down, and only quantum mechanics provides the concepts and computational methods that explain observed phenomena. Quantum mechanics uses specially developed tools, such as wave functions, that contain all of the relevant information about particles and systems. This approach is needed to understand the true nature of electromagnetic radiation and how that radiation interacts with matter. The mathematics used in quantum mechanics is highly specialized and often complex. In this chapter we'll address some of the key questions that require quantum mechanics without presenting the full mathematical formalism. The important questions we'll address all deal with observable physical phenomena. We'll discuss how quantum mechanics is used to answer these questions but will omit the advanced mathematical details.

WHAT IS QUANTIZATION?

Matter or energy is said to be **quantized** if it appears only in certain discrete sizes. The **quantization** of matter is suggested by chemical analysis. Some relevant results from chemistry are as follows:

- The law of definite proportions (or Proust's law) states that any chemical compound always contains the same proportions of elements by mass. For example, the proportion of oxygen to carbon by mass is 4:3 in carbon monoxide and 8:3 in carbon dioxide.
- The law of multiple proportions (or Dalton's law) concerns what happens when two elements join to form different compounds. In that case, a fixed mass of one element can combine with two different

masses of the other element. The law says that those two masses will be related in a simple whole-number ratio. For the case of carbon monoxide and carbon dioxide, a fixed mass of carbon (say, 150 g) can combine with 200 g of oxygen to make carbon monoxide or 400 g of oxygen to form carbon dioxide. The ratio of 400 g to 200 g is 2:1.

- Analysis of atomic masses shows that all atoms have a mass very close to an integer multiple of the mass of hydrogen, the lightest element. Different isotopes of the same element have different masses, but each is still roughly an integer multiple of hydrogen's mass.
- Chemical reactions involve the transfer of definite amounts of charge, each of which is an integer multiple of the smallest observed unit of charge. The smallest positive charge is that of an ionized hydrogen atom, and the smallest negative charge (equal in magnitude to the smallest positive charge) is that of an electron.

How Are Mass and Electric Charge Quantized?

In 1897 J. J. Thomson found more direct experimental evidence for the quantization of mass and charge. Thomson used an evacuated tube, in which he used a heated metal filament to generate a stream of electrons. The electron stream was then accelerated through a potential difference and deflected electromagnetically. By measuring the amount of deflection, Thomson determined that that charge-to-mass ratio for electrons is significantly greater than that of ionized hydrogen, the lightest element. (A more precise modern determination gives a charge-to-mass ratio of 9.6×10^7 C/kg for ionized hydrogen and 1.8×10^{11} C/kg for electrons, a factor of nearly 2000 times larger.) Because the net positive and negative charges in a neutral atom must be equal, its larger charge-to-mass ratio implies that the electron's mass is very small compared with the rest of the atom. The electron's mass is 9.11×10^{-31} kg, and its charge is -1.60×10^{-19} C. The magnitude of that charge (1.60×10^{-19} C) is given the symbol e. That quantity $e = 1.60 \times 10^{-19}$ C is the fundamental unit of charge in atoms. Chemical reactions—which involve exchanges of electrons—involve multiples of that charge. Physicists consider electrons to be fundamental particles—meaning that they cannot be divided into anything smaller.

Protons and neutrons, the particles that make up the atomic nucleus, are not fundamental. Each one is composed of three quarks (Chapter 8), which accounts for the overall mass and charge of the proton and neutron. The proton's charge is $+e$ ($= 1.60 \times 10^{-19}$ C), and the neutron's charge is zero. The proton's mass is about 1.6726×10^{-27} kg (or 1.0073 u), and the neutron's mass is only slightly more, about 1.6749×10^{-27} kg (1.0087 u).

The masses and electric charges of the proton, neutron, and electron explain the quantization observed in chemical reactions. For convenience, atomic masses are normally expressed in atomic mass units u, with 1 u = 1.6605 × 10^{-27} kg. That's close to the mass of the proton and neutron, so an atom consisting of Z protons and N neutrons has a mass very close to $Z + N$ expressed in atomic mass units. For example, common oxygen has eight protons and eight neutrons, and thus an atomic mass of about 16 u. With 92 protons and 136 neutrons, uranium has a mass of about 238 u. The masses are not exactly equal to a whole number of atomic mass units, due to an effect called binding energy (Chapter 7), but they are always close to that whole number. Electrons are not massive enough to contribute significantly to the mass expressed in u. These consistent whole-number masses are responsible for Proust's law and Dalton's law in chemistry.

In chemical reactions, electrons are gained or lost by atoms. (The positively charged protons are stuck in the nucleus, so they aren't transferred in chemical reactions.) Because each electron's charge is $-e$, the net charge involved in any reaction is an integer multiple of e. Atoms can be **ionized,** meaning they obtain a net charge by gaining or losing electrons. Sodium tends to lose a single electron to form the ion Na^+, while fluorine easily gains an electron to form Fl^-. Oxygen often gains two electrons to form O^{-2}. The affinity for gaining or losing electrons explains many observed chemical reactions, as well as the formation of chemical bonds—for example, in water (H_2O).

In addition to the protons, neutrons, and electrons that make up ordinary matter, there are other particles—some fundamental and others composed of various combinations of quarks. Most of these are short-lived and are created in nuclear or high-energy reactions. However, each of them has a well-defined mass and a charge that is either zero (like the neutron) or some multiple of the fundamental charge e. For example, a muon is a fundamental particle with charge $-e$ and mass 1.9×10^{-28} kg (more than an electron but less than a proton). A Σ^+ particle is composed of three quarks and has charge $+e$ and mass 2.0×10^{-27} kg (slightly more than a proton). See Chapter 8 for a more complete discussion of particles.

How Is Atomic Energy Quantized?

Fluorescent lights are one example of how we use atomic energy in everyday life. The closed tube contains a gas of one or more types of atoms or molecules. Mercury vapor is common in household lighting, although it's now being replaced by other gases for reasons of safety and to produce a more pleasing, "warmer" glow. (Pure mercury vapor gives off light with a bluish tint.) Neon lights give a distinctive bright red glow and are used mainly for outdoor displays.

Figure 4.1 shows the characteristic atomic spectra of two elements, hydrogen and helium. In each case the spectrum is produced in a gas of the pure element in a closed tube. When an electric potential is placed across the ends

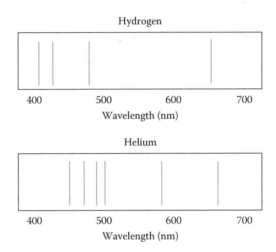

of the tube, atoms absorb some of the electrical energy. The atoms release the excess energy in the form of light. That light is sent through a diffraction grating, which separates it into different wavelengths. Figure 4.1 shows that each different kind of atom produces a unique set of visible wavelengths, which can be used to identify the elements present in a gas sample. For example, astronomers use this method—called **atomic spectroscopy**—to analyze light from distant stars and galaxies to determine their composition. The fact that all atomic spectra contain specific wavelengths rather than a continuous visible spectrum shows that atoms can only absorb and emit energy in specific amounts. That is, atomic energy is quantized.

What Is Blackbody Radiation?

Exciting atoms electrically is one of many ways to produce visible light. Light is also emitted when materials are heated to extremely high temperatures. You can see a piece of metal glowing when it's "red hot," and electric heating elements on stovetops glow red-orange when you select a high-temperature setting. Light (or other electromagnetic radiation, such as infrared radiation) emitted from hot objects is called **blackbody radiation.** That might seem like an odd name because a glowing body certainly doesn't appear black. The name comes from the fact that, in everyday situations, white objects reflect incoming light and black objects absorb it. Technically, a perfect blackbody is a perfect absorber of energy. That's what allows a blackbody to increase its temperature and then emit excess energy as electromagnetic radiation. A blackbody at a constant temperature is in equilibrium with its surroundings, absorbing and emitting energy at the same rate.

The spectrum of electromagnetic radiation emitted by a blackbody is continuous—not discrete like atomic spectra. Figure 4.2 shows that the spectrum has a peak at a particular wavelength λ_{max}, which depends on the blackbody's temperature. The relationship between peak wavelength λ_{max} and temperature is given by **Wien's law:**

$$\lambda_{max}T = 2.898 \times 10^{-3} \text{ m} \cdot \text{K} \tag{4.1}$$

The peak output of a blackbody is shifted to shorter wavelengths as the temperature increases. Your stovetop heating element turned to "low" emits infrared radiation that you can feel if you put your hand near it, but there's not enough visible radiation to see. Turn the setting to "high" and more of the spectrum will fall in the high-wavelength part of the visible spectrum. That's when you see it glowing red-orange.

There are other common examples of blackbody radiation. Humans are normally only slightly warmer than their surroundings, but we emit infrared radiation that can be detected by night-vision goggles. Your body temperature is about 310 K, for which Wien's law predicts a peak emission at a wavelength of about 0.01 mm in the infrared spectrum. If you look at the stars, you notice that some stars have a reddish tint and some other stars appear slightly blue. Wien's law tells you that the red ones are cooler and the blue ones are hotter. Our sun's temperature lies between those extremes, so its light is yellowish white.

A blackbody's radiation output is proportional to the fourth power of its absolute temperature. This fact is contained in the **Stefan–Boltzmann law,**

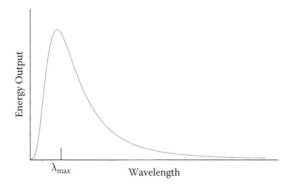

Figure 4.2 Radiation spectrum from a blackbody.

an expression for $R(T)$, the power per unit area emitted by a blackbody at temperature T:

$$R(T) = \varepsilon \sigma T^4 \qquad (4.2)$$

where $\sigma = 5.67 \times 10^{-8}$ W/(m$^2 \cdot$K^4) is a constant and ε is called the **emissivity** of the blackbody. Emissivity ε is a dimensionless number between 0 and 1, with 0 representing a perfect reflector and 1 a perfect blackbody. The Stefan–Boltzmann constant can be used to estimate the surface temperatures of bodies that approximate blackbodies, such as the sun.

How Is Electromagnetic Radiation Quantized?

In 1900 physicist Max Planck developed a theoretical model for how blackbody radiation works. Planck's model was based on the blackbody containing simple harmonic oscillators with different frequencies, f, absorbing energy from an oscillating electric field and emitting it as electromagnetic radiation. He found that the only way to produce the observed blackbody spectrum (Figure 4.2) was to assume that the emitted electromagnetic radiation is quantized in the form

$$E = hf \qquad (4.3)$$

where $h = 6.626 \times 10^{-34}$ J·s—now called **Planck's constant.** When a blackbody is in equilibrium with its surroundings, it also absorbs electromagnetic energy. Thus, the oscillators within the blackbody can only absorb or emit energy in multiples of hf. Planck's theoretical work was an indication that electromagnetic radiation is quantized in the form given by Equation (4.3).

What Is a Photon?

Equation (4.3) is not a result peculiar to blackbody radiation. Rather, it's a general expression that's valid for any electromagnetic wave of frequency f. All electromagnetic radiation is quantized in this way, and physicists view Planck's constant as one of the most important fundamental constants in nature. Inherent in this result is the idea that electromagnetic radiation is quantized in units hf. A single quantum of radiation with energy $E = hf$ is called a **photon,** regardless of whether the photon is visible or in any other part of the electromagnetic spectrum.

In vacuum, electromagnetic waves travel with speed $c = 2.998 \times 10^8$ m/s (Chapter 2). For waves with that speed, the frequency f and wavelength λ are related by $f = c/\lambda$. Therefore, a photon's energy is expressed in terms of its wavelength as

$$E = \frac{hc}{\lambda} \qquad (4.4)$$

For example, a 550-nm photon (in the middle of the visible spectrum) has energy 3.6×10^{-19} J in SI units. That's an extremely small amount of energy, which explains why you don't notice the quantization of light energy in everyday life. It takes a large number of photons before your eye can detect them, or before you notice a stream of infrared photons that warm your skin. However, the relationship between energy and wavelength given by Equation (4.4) has been verified experimentally throughout the electromagnetic spectrum, from gamma rays (with short wavelength and high energy) through radio waves (with long wavelength and low energy).

For visible light, such as the hydrogen and helium spectra shown in Figure 4.1, the inverse relationship between energy and wavelength means that a photon with a shorter wavelength has a higher energy than a photon with a longer wavelength. Thus, in the visible spectrum, violet photons have the most energy and red photons the least. And as we noted earlier, the fact that these atomic spectra contain specific wavelengths rather than a continuous distribution is evidence that atomic energy is quantized. An atom that undergoes a spontaneous transition from a higher to a lower energy state emits a photon with energy $E = hc/\lambda$ equal to the difference between the two atomic energy states (minus any recoil energy involved in the emission process).

Because of the small quantity of energy in single photons—on the order of 10^{-19} J for visible light—physicists often express photon energies in units of electron-volts, with 1 electron-volt (eV) defined as the energy gained by a single electron accelerated through a potential difference of 1 V. The conversion between joules and electron-volts is 1 eV $= 1.60 \times 10^{-19}$ J. Visible photons range in energy from 1.8 eV for 700 nm red to 3.1 eV for 400 nm violet. Outside the visible spectrum, photon energies stretch far in both directions, from less than 1 μeV for long radio waves to more than 1 MeV for gamma rays.

What Is the Photoelectric Effect?

Planck was originally unsure of his hypothesis, $E = hf$, and the implication of quantized energy. Although quantization explained the radiation spectrum from blackbodies, it seemed to Planck to be an ad hoc assumption that couldn't be justified by the fundamental physics known at the time.

In 1905 Albert Einstein provided further support for the quantum hypothesis by using it to explain the **photoelectric effect,** a phenomenon first observed in the 1880s by Heinrich Hertz. As its name implies, the photoelectric effect occurs when light—usually visible or ultraviolet—strikes a metal surface, causing electrons to be ejected by the metal. If Planck's quantum hypothesis is correct, then Equation (4.3) implies that the ejected electrons (called photoelectrons) should have kinetic energies that vary depending on the frequency (or wavelength) of the incident light.

An experiment to test this idea is shown schematically in Figure 4.3. Within an evacuated glass tube, photoelectrons are emitted by one metal and collected by another, with the photoelectron rate measured as an electric current by an ammeter. The power supply is set up with a polarity that pushes electrons away from the collector. When the power supply voltage is increased gradually from zero, it will reach some voltage V_0, called the stopping potential, just as the photoelectron current reaches zero. Thus, the stopping potential effectively measures the maximum kinetic energy K_{max} of the photoelectrons, with $K_{max} = eV_0$.

When the experiment is performed by recording the stopping potential V_0 as a function of the incident light's frequency f, the results are as shown in Figure 4.4. The graph of eV_0 versus frequency is a straight line, with an intercept at some energy $-W$. Einstein explained the results by starting with Planck's assumption that light is quantized. A single photon with energy $E = hf$ strikes an electron in the metal, transferring some or all of its energy to the electron. If all the photon's energy is transferred, electrons with a maximum kinetic energy of

$$K_{max} = eV_0 = hf - W \tag{4.5}$$

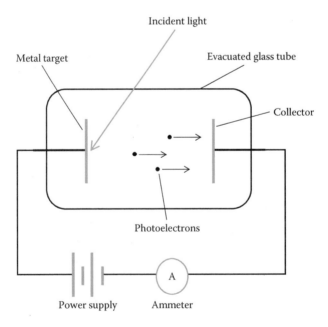

Figure 4.3   Device used to study the photoelectric effect.

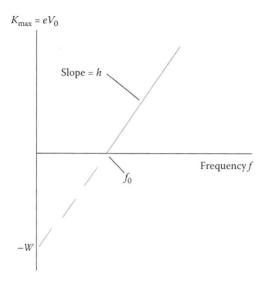

Figure 4.4   Results of a photoelectric effect experiment.

are produced, with K_{max} equal to the difference between the incoming photon energy, hf, and the metal's work function W, the small amount of energy that binds electrons to the metal. Equation (4.5) is consistent with the experimental results in Figure 4.4, which shows a straight line with slope h and intercept $-W$.

The photoelectric effect gives a way to measure Planck's constant h, which is just the slope of the graph in Figure 4.4. Measuring Planck's constant in this way gives numerical results (with modern value $h = 6.626 \times 10^{-34}$ J·s) that are consistent with the value of h needed to explain the blackbody spectrum (Figure 4.2). Performing the experiment with different metal targets measures the work function of different metals, such as 4.1 eV for aluminum and 5.0 eV for nickel.

Notably, the photoelectron energy depends only on the incident light's frequency (or wavelength), rather than on the light's intensity (brightness). This is evidence that photoelectrons are not produced by the incoming light heating the surface, but rather by individual photons with energy hf dislodging a single electron at a time, consistent with Figure 4.4. Further, there is a threshold frequency f_0, below which no photoelectrons can be produced because the incoming photons have insufficient energy to overcome the target metal's work function. The results all support Planck's assumption that light is quantized as photons of energy hf. It's now a well-established fact that light is carried by photons with energy $E = hf$.

WHAT IS WAVE–PARTICLE DUALITY?

Light was established as an electromagnetic wave by the late nineteenth century. It's impossible to understand many optical phenomena—particularly diffraction, interference, and polarization—without treating light as a wave. However, other phenomena, such as blackbody radiation and the photoelectric effect, can only be understood if light is modeled as a stream of photons—small quanta of energy that are so localized in space that they can be thought of as particles. This sets up an apparent contradiction. How can light be both a wave and a particle? Waves and particles are very different things, conceptually. If both wave and particle aspects are needed, it's hard to imagine how wave and particle properties might coexist in a photon that appears to be much smaller than a single wavelength of light. Nevertheless, because light clearly has a dual nature, we must accept the resulting **wave–particle duality.**

What Does the Two-Slit Experiment Reveal about Wave–Particle Duality?

One of the clearest manifestations of light's wave nature is the two-slit interference experiment, also commonly known as the double-slit experiment (Figure 4.5). In this experiment coherent light (for example, from a laser) is shined though a pair of parallel, closely spaced slits in an otherwise opaque material. Light emerging from the slits is diffracted (bent), so that light from each slit can overlap and interfere with light from the other slit. The result is a distinct pattern of constructive and destructive interference, manifested as light and dark areas on a distant viewing screen. The results are well understood in terms of wave interference.

The same experimental arrangement can also reveal light's particle (photon) nature. Because individual photons carry very little energy (Equation 4.4), the light source's intensity can be turned down to a very low level so that, effectively, one photon at a time passes through the double slits. Doing the experiment this way allows you to see the interference pattern forming gradually (Figure 4.6) and at the same time shows light's photon nature. After a long enough time has elapsed, the familiar interference pattern is formed. (Also see "How Does Interference Reveal Light's Wave Properties?" in Chapter 5.) Although you can see individual photons reaching the screen, the interference pattern shows that the photons still carry wave information.

However, if the light intensity is so low that photons pass through the slits sequentially, separated from one another, how can interference result? In that case, just what is interfering with what? The only possible explanation is that each photon must go through both slits at once. This fact is difficult to

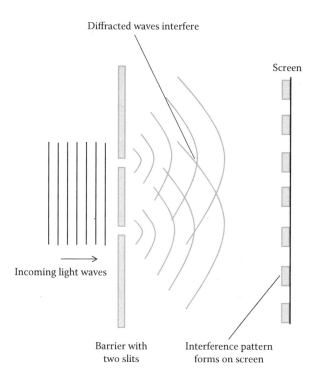

Diffracted waves interfere

Screen

Incoming light waves

Barrier with
two slits

Interference pattern
forms on screen

Figure 4.5 The double-slit experiment.

reconcile with the results of classical physics, which is based on observation of macroscopic objects. Classically, the particle-like photon must pass through one slit or the other, not both. This experiment is often cited as an example of how quantum physics differs from our everyday experience.

In another variation of this experiment, photon detectors are placed just past each slit. The idea here is to detect and count the single photons as they pass through the pair of slits, in order to remove the ambiguity over which slit an individual photon passes through. When the experiment is performed this way, individual photons can be counted by each detector and thus identified with passing through one slit or the other, not both. However, in doing so the two-slit interference pattern disappears. Adding the photon detectors to the apparatus actually changes the observed outcome.

Two important points arise from the photon-counter version of the two-slit experiment. First, it illustrates that, on the quantum level, the interaction between experimenter and experiment may have a considerable effect on the outcome. Scientists or their equipment can't be thought of as passive observers of nature; rather, they must be considered as active participants in a physical

Time

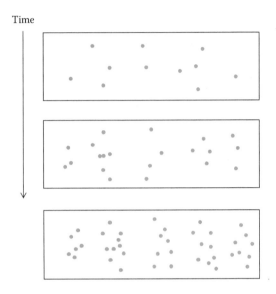

Figure 4.6   Time sequence of a double-slit experiment with low light intensity.

process. The second point is that although waves and particles are both fundamental features of light, they can't be observed simultaneously. In this case, the
photon counters revealed light's particle nature, and in doing so they made it
impossible to see wave interference. Niels Bohr explained that light's wave and
particle properties are **complementary,** meaning that they are both essential
features but mutually exclusive in that they can't be observed simultaneously.
Bohr and his colleagues extended the principle of complementarity to other
physical phenomena in the quantum world.

What Are Particle Waves?

Wave–particle duality is not restricted to light. Discrete particles such as electrons or protons, as well as whole atoms and molecules, also exhibit wave properties. In general, a particle with momentum p has wavelength

$$\lambda = h/p \qquad (4.6)$$

Equation (4.6) is called the **de Broglie relation,** after Louis de Broglie,
who proposed it in 1924. At that time there was no experimental evidence of
particles displaying wave properties, but soon after, this constructive wave
interference was seen in electrons scattered from solid targets. Analysis of
such experiments shows that the interference results when electrons scatter

from adjacent layers of atoms in the solid's lattice, with measured wavelengths consistent with Equation (4.6).

The wave nature of particles is quite real and can be seen in any experiment that will reveal wave properties. For example, a stream of electrons sent through a pair of parallel slits will produce an interference pattern like the one shown in Figure 4.5. The small numerical value of Planck's constant generally makes particle wavelengths very short, and because the wavelength is inversely proportional to momentum, particle wavelengths are easiest to measure using subatomic particles, which have small masses. For that reason, **particle waves** are viewed as a quantum effect, even though in principle any particle with momentum has a wavelength.

A widely used application of the wave properties of electrons is in the **electron microscope.** Even with finely ground lenses and computer-aided imaging techniques, optical microscopes are limited by visible light's wavelength range of 400–700 nm. Objects smaller than that size can't be resolved in an optical microscope. In an electron microscope, imaging is done with a beam of electrons reflected from or transmitted through the object of study. According to the de Broglie relation, faster electrons have shorter wavelengths, and the electron beam can be tuned to the resolution desired. The best electron microscopes can resolve detail down to 10ths of nanometers—essentially the size of single atoms.

Describing particles as waves is an important conceptual step toward building the full quantum theory we use today. In the mathematical formalism of quantum mechanics, each particle has an associated **wave function** that contains important physical information, such as position and momentum. Applying appropriate mathematical operations allows that information to be extracted and interpreted physically.

WHAT IS THE HEISENBERG UNCERTAINTY PRINCIPLE?

When it comes to making precise measurements in the quantum world, the determination of which slit a photon or electron passes through in a two-slit experiment is just the tip of the iceberg. Whenever you make any physical measurement, you're used to dealing with some uncertainty. For example, suppose you step on a digital scale that reads your weight to 10ths of a pound. If the scale is a good one, then you know your weight to within 1/10 of a pound, but no better. Use a ruler marked in millimeters to measure the width of a piece of paper. If you're careful, you can surely measure the width to the closest millimeter and maybe some fraction thereof. But there's no way you could measure to 0.01 mm or better with any confidence. In the world of classical physics, you ultimately come up against some limit of precision, based on limitations of the equipment or your ability to see in fine enough detail.

Such limitations also exist in measurements made in the quantum world, but there's a further, more stringent restriction on precision. Particles exhibit wave behavior, and waves can't be localized to a point in space. A good model for this is a **wave packet,** which contains an assortment of wavelengths spread out over space. By analyzing the behavior of wave packets, Werner Heisenberg showed that there are theoretical limits to what can be known about the wave's properties independent of the precision of the instrument used to measure any particular quantity.

For example, for a particle moving in one dimension (x), the uncertainty Δp in its momentum and the uncertainty Δx in its position are related by the **Heisenberg uncertainty principle:**

$$\Delta p \Delta x \geq \hbar/2 \tag{4.7}$$

where $\hbar = h/2\pi$. (The symbol $\hbar$ is used for convenience because the combination $h/2\pi$ appears frequently.) The uncertainty principle places a limit on the product of uncertainties of momentum and position, so a more precise measurement of one implies less precision in the other. This helps us interpret the results of the double-slit experiment with single photons (or electrons). For one thing, it's impossible to track a single photon's precise path from the light source through the slits to the observing screen. Further, inserting photon counters to measure a photon's position to greater precision implies less precise knowledge of the photon's momentum (and wavelength, according to the de Broglie relation), which corresponds to the disappearance of the interference pattern.

Other forms of the Heisenberg uncertainty principle connect some other pairs of physical quantities that have a product with the same units as Planck's constant. For example, energy and time are two such related quantities, with

$$\Delta E \Delta t \geq \hbar/2 \tag{4.8}$$

Equation (4.8) is often applied by physicists when they study short-lived excited atomic or nuclear states. For states with extremely short lifetimes, there is a correspondingly large ΔE, or a lack of precision in measuring that state's energy.

 ## WHAT DOES QUANTUM MECHANICS TELL US ABOUT HYDROGEN ATOMS?

A simple but important and exemplary system that can be solved exactly using quantum mechanics is the hydrogen atom. Hydrogen has a single proton (electric charge $+e$) and a single electron (electric charge $-e$) separated by distance r. This makes it possible to write the potential energy as a simple

GOING DEEPER—WAVE FUNCTION PROBABILITY
AND THE SCHRÖDINGER EQUATION

A particle in one-dimensional motion (x) is described by a wave function $\psi(x)$ that contains physical information about the particle. In general, $\psi(x)$ is a complex function, meaning that it has parts that are real and parts that are imaginary, containing the number $i = \sqrt{-1}$. Multiplying a complex number by its complex conjugate (indicated by *) results in a real number. For the wave function $\psi(x)$, that product is

$$[P(x)]^2 = \Psi^*(x)\Psi(x) \tag{4.9}$$

with $P(x)$ equal to the probability of finding the particle as a function of its position x. The concept of measuring a particle's position as a distribution of probabilities is consistent in a conceptual sense with the uncertainty principle's restriction on measuring the particle's position precisely.

Other information contained in the wave function can be extracted by the use of **operators.** An example of an operator (designated with the carat symbol ^) is the one for momentum

$$\hat{p} = -i\hbar \frac{d}{dx}$$

When the momentum operator acts on the wave function $\psi(x)$, the result is the measurable momentum p multiplied by the wave function, or

$$\hat{p}\Psi = -i\hbar \frac{d\Psi}{dx} = p\Psi \tag{4.10}$$

In a time-independent system with conservative forces, the total energy E of a particle is the sum of kinetic and potential energy, or $K + U = E$. With $K = p^2/2m$, applying the momentum operator (Equation 4.10) gives

$$-\frac{\hbar^2}{2m}\frac{d^2\Psi}{dx^2} + U\Psi = E\Psi \tag{4.11}$$

Equation (4.11) is called the **time-independent Schrödinger equation** and can be used to solve for a particle's energy, provided the potential energy function U and boundary conditions are known.

For time-dependent systems, the wave function $\psi(x,t)$ depends on position and time, and the energy is no longer constant but rather is associated with the Hamiltonian operator

$$\hat{H} = i\hbar\frac{\partial}{\partial t}$$

The **time-dependent Schrödinger equation** uses the Hamiltonian to give the energy, and because the wave function depends on position and time, partial derivatives are required:

$$-\frac{\hbar^2}{2m}\frac{\partial^2 \Psi}{\partial x^2} + U\Psi = i\hbar\frac{\partial \Psi}{\partial t} \qquad (4.12)$$

Both the time-dependent and time-independent equations can be extended to three dimensions by including derivatives with respect to x, y, and z, or an appropriate set of three coordinates. In both time-dependent and time-independent formulations, solving the Schrödinger equation yields quantized wave functions and energy states.

function $U = -ke^2/r$, which can be used in a three-dimensional version of the time-independent Schrödinger equation (4.11).

Solving the Schrödinger equation for hydrogen yields quantized energy levels

$$E_n = -\frac{E_0}{n^2} \qquad (4.13)$$

where n is a positive integer (1, 2, 3,...) and $E_0 = 13.6$ eV. This result is consistent with the semiclassical Bohr result (Chapter 7) and with spectroscopic data, not only for the familiar visible spectrum (hydrogen in Figure 4.1) but also throughout the ultraviolet and infrared spectral regions, thus confirming the validity of the quantum-mechanical approach.

The complete solution to the Schrödinger equation for the hydrogen atom also tells us the allowed wave functions for the electron. Those wave functions depend on three **quantum numbers,** which arise naturally from the boundary conditions needed for realistic solutions. The three quantum numbers and the restrictions on them that arise from the boundary conditions are

n (principal quantum number) = 1, 2, 3,...
l (orbital angular momentum quantum number) = 0, 1, 2,...up to $n - 1$
m_l (magnetic quantum number) = 0, ±1, ±2,...with $|m_l| \leq l$

Each set of three allowed quantum numbers defines an allowed state for the atom. Equation (4.13) shows that states with lower n quantum numbers are lower in energy, so the state with the lowest energy has quantum numbers (n, l, m_l) equal to $(1, 0, 0)$. (Note how the restrictions on quantum numbers require both l and m_l to be equal to 0 for $n = 1$.) For $n = 2$ there are four possible (n, l, m_l) states: $(2, 0, 0)$, $(2, 1, 0)$, $(2, 1, 1)$, and $(2, 1, -1)$. In general, a given energy level with quantum number n has a total of n^2 possible (n, l, m_l) states. Many of an atom's important characteristics depend on only n and l, so a common shorthand notation is to give just a state's n and l quantum numbers using the number for n and a letter for l: s for $l = 0$, p for $l = 1$, d for $l = 2$, and f for $l = 3$. For example, the state $(1, 0, 0)$ is called 1s, $(3, 1, 1)$ is 3p, and $(4, 3, -2)$ is 4f.

Each hydrogen wave function $\psi(n, l, m_l)$ is a function of the three spatial dimensions. Just as in one dimension (Equation 4.9), $\psi^*\psi$ is the square of the spatial probability distribution. For the hydrogen atom, $\psi^*\psi$ reveals the probability distribution for the electron in space, effectively showing the atom's shape. A few of these distributions are shown in Figure 4.7. Notice that s states are spherically symmetric, while p states tend to have higher probability densities along one axis when $m_l = 0$ and perpendicular to that axis when $m_l = \pm 1$. With more possible combinations of quantum numbers, d and f states show wider variety of spatial distributions.

How Does Quantum Mechanics Describe Transitions between States?

The electron in hydrogen can make transitions between states. If there is an external source of energy that the atom can absorb, this forces the electron to a state with higher energy. If the electron is in any state above the $n = 1$ **ground state,** it can make a spontaneous transition to a lower state, giving up the energy difference in the form of a photon. That's the source of the observed hydrogen spectrum (Figure 4.1).

The hydrogen wave functions and corresponding quantum numbers govern which transitions are **allowed** and which are **forbidden.** There is no restriction on how n changes in a transition. However, the quantum number l can only change by ± 1. Therefore, an atom in a 3p state may drop spontaneously to a 2s or 1s state but not to 2p. Similarly, the magnetic quantum number m_l may only change by either 0 or ± 1. This means that a transition from $(3, 2, 2)$ to $(2, 1, 1)$ is allowed, but transitions to other 2p states (with $m_l = 0$ or -1) are forbidden.

How Does Quantum Mechanics Apply to Other Atoms?

The next atom in the periodic table above hydrogen is helium, with nuclear charge $+2e$ and two electrons. Adding the second electron makes for a much

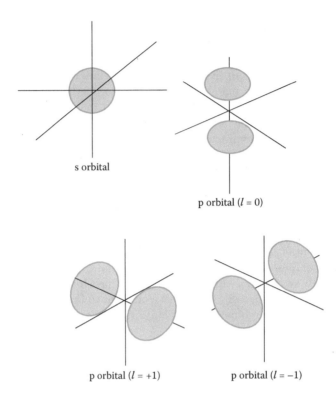

s orbital

p orbital ($l = 0$)

p orbital ($l = +1$)

p orbital ($l = -1$)

Figure 4.7   Probability distributions for some hydrogen states.

more complicated system because, in addition to each electron's attraction to the nucleus, there is the electron–electron repulsion to consider. This makes it impossible to solve the Schrödinger equation exactly for helium. Some approximations give reasonable results for wave functions and energy levels, but there's no simple closed-form solution analogous to Equation (4.13). For larger atoms, the mathematics is even more complex, and it's impossible to find even approximate solutions.

However, many of the principles that are firmly established from the exact quantum-mechanical solution to hydrogen can still be applied to the study of other atoms. For one thing, characteristic spectra (such as the helium spectrum in Figure 4.1) reveal the existence of quantized energy levels. The irregularity of helium's spectral lines corresponds to the fact that no simple relationship gives all the energy levels as a function of quantum numbers. The same is true for atoms larger than helium, which have their own unique spectra. The evidence of quantized energy levels is confirmed by absorption spectra, which show that atoms absorb energy in the same quantized chunks in which they emit energy.

Many properties of atoms throughout the periodic table can be understood with the quantum numbers introduced for hydrogen (n, l, m_l), plus a fourth quantum number, m_s, that's due to the electron's intrinsic angular momentum. The two allowed values of m_s are $\pm\frac{1}{2}$, independently of the other three quantum numbers. The classical analog to intrinsic angular momentum is a spinning object, so the two values of m_s are often referred to as "spin $\frac{1}{2}$" and "spin $-\frac{1}{2}$."

A key to understanding larger atoms is the **Pauli exclusion principle,** which states that no two electrons in an atom can have the same set of four quantum numbers (n, l, m_l, m_s). This rule, combined with the restrictions on allowed combinations of n, l, and m_l established for hydrogen, limits the allowed quantum numbers for electrons in a multielectron atom. For example, an atom can have only two electrons with $n = 1$, with quantum numbers $(1, 0, 0, +\frac{1}{2})$ and $(1, 0, 0, -\frac{1}{2})$. Similarly, for $n = 2$, there are two possible sets of quantum numbers for a 2s electron and six for a 2p electron, making eight total states with $n = 2$. The number of states increases with n and is equal to $2n^2$.

The four quantum numbers, coupled with the Pauli principle, provide the physical basis for the empirical rules that are used to organize the periodic table of elements. A group of electrons with the same n forms a **shell,** and a group with the same nl combination (for example, 2p or 4d) forms a **subshell.** The ground state of a multielectron atom is the one in which the electrons are organized to have the least possible energy, so think of a multielectron filling its subshells from lowest to highest energy, in order. Similarly to hydrogen, lower values of n generally correspond to lower energy, but there are some exceptions. The first two electrons go into the 1s subshell, followed by two more electrons into 2s and six electrons in 2p. A more complete listing of the order is

1s, 2s, 2p, 3s, 3p, 4s, 3d, 4p, 5s, 4d, 5p, 6s, 4f, 5d, 6p, 7s, 5f, 6d, and 7p

The physical properties and chemical behavior of atoms are understood in part by knowing the atom's subshell configuration and, in particular, the number of electrons in the highest unfilled subshell. Alkali metals such as sodium and potassium have a single electron in an s subshell, which they readily give up to form positive ions. Halogens such as fluorine and chlorine have five electrons in an unfilled p subshell, and they tend to add an electron to fill the subshell, thus becoming a negative ion. Alkali metals and halogens pair to form salts, including common NaCl. Copper, silver, and gold (in the same column, or group, of the periodic table) each has a single s electron that is bound very weakly. These metals are excellent electrical conductors because even a weak applied electric field will make the electrons move, resulting in electric current. Noble gases (including common gases helium and argon) have filled s and p orbitals. This makes them extremely stable and resistant to chemical bonding, even with each other, so they are monatomic, have low boiling points, and rarely participate in chemical reactions.

The shapes of unfilled electron orbitals (Figure 4.6) can be used to explain an atom's behavior in more detail. A well-known example is carbon, with two p electrons in an unfilled subshell. Organic chemists use the shapes of the p orbitals in carbon to explain the structures and properties of many organic molecules. Having just two p electrons allows carbon a wide range of bond types, including bonding to other carbon atoms via single or double bonds. Pure carbon forms two well-known distinctive structures. In its two-dimensional form, pure carbon is known as graphite—the stuff in your pencil—with the two-dimensional structure allowing layers of atoms to slide off easily. The three-dimensional form of carbon is diamond, one of the strongest natural materials known because of the strength of the bonds between carbon atoms. In recent years several new pure carbon structures have been discovered. These include the structures shown in Figure 3.3, as well as **graphene**, with carbon atoms arranged in a hexagonal lattice that is also extremely strong. Graphene has unusual properties of electrical conduction that may have useful applications in electronic devices.

What Is the Zeeman Effect?

Under normal conditions, the energy of a hydrogen atom is determined for all practical purposes by its principal quantum number n (Equation 4.13) independently of the other quantum numbers. (This is an example of what physicists call **degeneracy,** where multiple states have the same energy.) The situation changes if the sample of hydrogen is put into a magnetic field. Because of the electron's orbital angular momentum, it has a magnetic moment proportional to the quantum number m_l. The magnetic field interacts with the magnetic moment, splitting the energy levels by an energy proportional to m_l and to the applied field. This is the **Zeeman effect** and it's observed by the resulting splitting of a single spectral line into three equally spaced lines (Figure 4.8). The Zeeman effect provides tangible evidence of the electron's orbital magnetic moment and proves that each atomic state has an associated m_l value. One application of the Zeeman effect is in astrophysics, where the observed splitting of spectral lines indicates the strength of the internal magnetic field in stars.

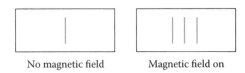

No magnetic field   Magnetic field on

Figure 4.8   Splitting of a hydrogen spectral line by the Zeeman effect.

What Is the Stern–Gerlach Experiment?

In the **Stern–Gerlach experiment,** a beam of neutral atoms is sent through an inhomogeneous magnetic field. An inhomogeneous field exerts different forces on atoms having different magnetic moments, separating them physically from one another. The experiment was originally done with silver atoms, but it can be done with any atom having zero orbital magnetic moment, so that there will be no Zeeman-type splitting. The result is a separation of the atoms into two distinct groups, associated with the two spin states $m_s = \pm\frac{1}{2}$. This experiment provides evidence of the two spin states and the electron's intrinsic angular momentum.

What Are NMR and MRI?

A similar splitting of energy levels in the presence of an applied magnetic field occurs within the nucleus because, like electrons, protons and neutrons also have intrinsic magnetic moments. This is the basic idea behind **nuclear magnetic resonance** (**NMR**). A nucleus (often a hydrogen nucleus, which is a single proton) has multiple spin states that have slightly different energies in the presence of a magnetic field. The splitting is small, corresponding to the energy of a radio-frequency photon. A maximum absorption—or resonance—of those photons occurs when their energy matches the transition energy between spin states. Because the energy difference also depends on the nucleus's local environment, an absorption spectrum reveals details about the material's composition and structure.

Essentially, the same technique is used in **magnetic resonance imaging** (**MRI**) of the human body. The part of the body being imaged is placed in a strong magnetic field, which aligns the protons in water molecules (H_2O) within the body. Then a radio-frequency field is applied, causing some of the aligned magnetic moments to flip. When the RF field is turned off and the moments return to their former state, the emitted energy is detected and, with computer analysis, is used to generate detailed images from inside the body. The method works because different body tissues carry different amounts of water. MRI is particularly useful in spotting irregularities in structure such as tumors, cardiovascular problems, and injuries to muscles, tendons, or ligaments.

WHAT IS QUANTUM TUNNELING?

Quantum tunneling occurs when a particle (or photon) travels through a region that's classically forbidden to it. This is possible because wave functions are not localized to a point in space, resulting in a nonzero probability of the particle appearing on the other side of the forbidden region. A simple example is a beam of light passing through glass. When the light strikes an interface with air at a steep angle, it suffers total internal reflection and can't pass into the air. However, if a second piece of glass is brought fairly close to the first, some of the light is transmitted into the second piece.

Another example is in electronics, where electrons can tunnel through an insulating barrier from one conductor to another. The amount of transmission depends on the electron's wave properties and the thickness of the insulator. A successful model of alpha decay from the nucleus (Chapter 7) uses the quantum tunneling concept. The alpha particle is trapped by the strong nuclear force but can tunnel through a relatively thin barrier at the nuclear surface. The tunneling probability, which can be computed in this model, is related directly to the half-life of nuclear decay.

 ## WHAT IS A QUANTUM COMPUTER?

In a modern (classical) computer, each bit of information is stored digitally as a 0 or 1. If you think about shrinking the size of the bit to a single atom, you could define the bit with, for example, two allowed spins: $+\frac{1}{2}$ and $-\frac{1}{2}$. However, a quantum-mechanical description of the atom might well be a **superposition** of the spin two states (i.e., some combination of them). A quantum computer uses superpositions of spin or some other property to define **qubits,** or quantum bits of information. The superposition allows much more information to be stored in a collection of qubits than in a similarly sized collection of definite 0 and 1 bits. As of today, only a few small prototypes of quantum computers have been made. There's still much research to be done on the most effective ways to store and process quantum information.

 ## WHAT ARE SOME OTHER APPLICATIONS OF QUANTUM MECHANICS?

In the twenty-first century, quantum mechanics is fully integrated into physics, so you now see quantum-mechanical concepts used routinely in a wide variety of ways. Accordingly, you'll see ideas from quantum mechanics show up in other chapters. This includes quantum statistics, which are needed to understand the behavior of some important thermodynamic systems (Chapter 6). In Chapter 7 more questions about atoms and nuclei will be considered, and these necessarily involve quantum concepts. Understanding fundamental particles (Chapter 8) requires quantum concepts, including some new ones not introduced in this chapter.

FURTHER READINGS

Ford, Kenneth J., and Goldstein, Diane 2005. *The Quantum World: Quantum Physics for Everyone.* Cambridge, MA: Harvard University Press.

Griffiths, David J. 2005. *Introduction to Quantum Mechanics,* 2nd ed. Upper Saddle River, NJ: Pearson Prentice Hall.

Thornton, Stephen T., and Rex, Andrew 2013. *Modern Physics for Scientists and Engineers,* 4th ed. Boston: Cengage Learning.

Chapter 5

Light and Optics

Light is a fundamental part of our lives. We use it in many different ways every day. Physics has a lot to tell us about light's behavior and how it's used both in the natural world and in applications that we've crafted to make our lives better. After reviewing some facts about light from earlier chapters, we'll address some important questions from optics and explain how we manipulate light by reflecting, bending, and focusing it. Understanding light's fundamental properties shows us how many natural phenomena work, such as rainbows and sparkling diamonds, and even the complex optical system in our own eyes. Over the centuries scientists have developed tools to see more than our eyes can, from microscopes that can focus on small cells to telescopes that let us see the outer parts of the universe. In modern times physics has brought us new ways to make and use light, such as lasers, optical fibers, and light-emitting diodes, and we'll finish the chapter by addressing these new tools and some applications that depend on them.

 WHAT IS LIGHT?

This apparently simple question has taken scientists many centuries to unravel. You've seen that light can be described as an electromagnetic wave (Chapter 2). What we call visible light covers the range of wavelengths from 400 to 700 nm—a fairly narrow part of the wide electromagnetic spectrum that runs from very short gamma rays ($\lambda < 10^{-10}$ m) to long radio waves ($\lambda > 1$ m). The colors of the visible spectrum are (in order of increasing wavelength) violet, blue, green, yellow, orange, and red. The observed speed of light (in vacuum), $c = 3.0 \times 10^8$ m/s, follows directly from electromagnetic theory. Modeling light as a wave explains phenomena such as diffraction and interference, which we'll address later in this chapter.

You've also seen that the electromagnetic wave alone is insufficient to describe how light behaves (Chapter 4). Rather, light consists of small quanta of energy called photons. In carefully designed experiments, individual photons can be detected and counted. A complete view of light includes both waves and particle-like photons, and we are left to contemplate the resulting wave–particle duality.

What Are Ray Optics and Wave Optics?

The dual nature of light suggests two ways to approach the study of optics. As the name suggests, **ray optics** (also called geometrical optics) assumes that light travels in a straight line until it encounters an object or change of medium, at which point the ray's change of direction can be predicted using systematic rules. For example, a ray of light (say, from your flashlight) that strikes a plane mirror reflects in a predictable way. The bending of light rays entering your eye focuses those rays on your retina and allows you to see. **Wave optics** (also called physical optics) is required whenever light exhibits interference effects, diffraction, or polarization. We'll explain all those effects in this chapter, after first considering some of the effects of ray optics—principally, reflection and refraction.

WHAT IS THE LAW OF REFLECTION?

The law of reflection says that a ray of light that strikes a mirror reflects at an angle equal to the angle made by the incident ray, as shown in Figure 5.1.

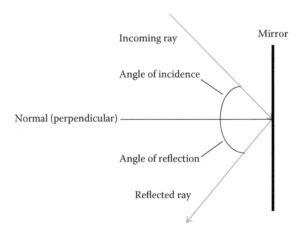

Figure 5.1   The law of reflection from a mirrored surface.

Reflections from any kind of mirrored surface follow from this simple rule. Notice that the angles of incidence and reflection are always measured with respect to a line called the **normal,** meaning that line is perpendicular to the surface at the point of incidence.

How Does the Law of Reflection Explain the Images You See in a Mirror?

What do you see when you look into your bathroom mirror every morning? First, note that your mirror is (probably!) flat. This is called a **plane mirror** because the flat surface lies in a single plane. Look at the reflection of your hand in a plane mirror. Based on the law of reflection, it can be shown that the image you see in the mirror is (1) the same size as the real hand, and (2) located on the opposite side of the mirror, the same distance from the mirror as your hand is from the mirror. The reflection looks the same in every respect, except that it's reversed so that a right hand looks like a left hand. That's because every part of the hand—for example, the thumb—has an image of itself directly across the mirror. Everything you see in a plane mirror is similarly reversed. A mole on your left cheek appears to be on your right cheek in the mirror image.

The image in a plane mirror is called a **virtual image** because the light rays don't actually reach the place where the image appears to be. You'll see later that some curved mirrors and lenses can make **real images,** where light rays from an object converge to form an image in space.

Why Doesn't Everything Reflect like a Mirror?

If you hold a mirror below your eyes, you can see an image of an overhead light. But if you look down onto the surface of your wooden desk, you don't see an image. The wood is still reflecting the light from above—that's why you can see it. However, most materials (like wood) exhibit **diffuse reflection,** where light striking it reflects in many directions and doesn't follow the law of reflection. When objects reflect diffusely, the light you see is reflected, but not from any one particular direction.

Many objects reflect not only diffusely but also selectively (i.e., not the same for all visible wavelengths). The leaves on the tree outside your window appear green because they are good reflectors of the green part of the visible spectrum and better absorbers of other colors.

How Do Curved Mirrors Form Images?

Light reflected from a curved mirror also obeys the law of reflection, but the curved shape means you won't see the exact copy you get in a plane mirror.

In you look at yourself in a **concave mirror,** the image of you that's formed depends on how close you are to the mirror. If you are close enough, the curve distorts your reflection and makes an enlarged, upright virtual image. But if you move back past a certain point, called the mirror's **focal point,** the reflected rays can flip over and form an inverted, real image that is enlarged if you're close to the focal point but reduced in size if you're farther away. You might have a small concave mirror in your bathroom and use it to get a better view of facial features (Figure 5.2). Just watch out for those larger-than-life blemishes!

Because their shape reflects rays outward rather than inward, **convex mirrors** always form virtual images that are reduced in size. Despite the reduced size, convex mirrors are very useful because the shape allows you to see a wider field of view (Figure 5.3). You see convex mirrors mounted in many places, indoors and out. Convex mirrors are placed at street intersections to help drivers see around the corner. You also see them in retail stores, to help merchants see shoplifters. Your outside rear-view car mirrors are slightly convex to give you a wider view, and the reduced image size leads to the common warning: Objects in mirror are closer than they appear.

If you don't want to buy a polished concave or convex mirror, just pick up a metal spoon. One surface is concave and the other is convex, and you can use it to try the effects we've been describing.

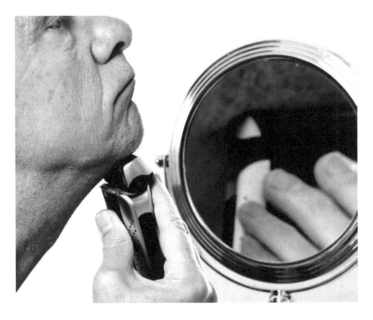

Figure 5.2   Enlarged image of an object placed inside the focal point of a concave mirror.

Figure 5.3   Image in a convex mirror.

WHAT IS REFRACTION?

Have you ever tried looking at a fish or some other object underwater? Your view is distorted because light rays coming from under the water's surface change direction when they enter the air. This is an example of **refraction,** which is the change in direction of a light ray as it goes from one medium to another. As shown in Figure 5.4, a ray of light that encounters a change in medium is partially reflected, as if the boundary were a mirror. The part of the ray that goes through is bent, from an angle θ_1 in medium 1 to angle θ_2 in medium 2. (Note that incident and refracted angles are measured with respect to the normal.) The two angles are related through a quantity known as **index of refraction** (symbol n), with

$$n_1 \sin\theta_1 = n_2 \sin\theta_2$$

(5.1)

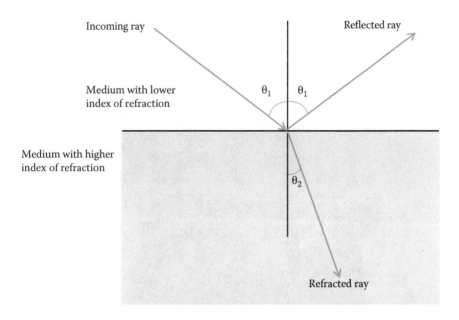

Incoming ray

Reflected ray

Medium with lower
index of refraction

θ_1 θ_1

Medium with higher
index of refraction

θ_2

Refracted ray

Figure 5.4   Reflection and refraction at a boundary between two media.

Equation (5.1) is known as Snell's law. For a vacuum, $n = 1$ by definition. In air, it's not much different: $n = 1.003$. But in water, $n = 1.33$, and in glass (depending on the type of glass), n is normally about 1.5. If you know the index of refraction of different media, you can use Snell's law to predict light's path in those media. For example, a light ray coming from that fish underwater that strikes the water's surface at an angle of 30° from the normal emerges in the air at about 42°. That's why the fish isn't where it appears to be!

Generally, denser materials have larger indexes of refraction, as is the pattern for air, water, and glass. However, there are exceptions. Modern plastics have been engineered to have a higher index of refraction. That's why your new plastic eyeglasses, with an index of refraction close to 1.7, don't need to be as thick as old-style glass lenses.

The index of refraction is closely related to the speed of light in a medium. In a vacuum, light travels at $c = 3.0 \times 10^8$ m/s. That's the maximum speed for light, and in any other medium it travels slower, with the speed v given by

$$v = c/n \tag{5.2}$$

where n is the index of refraction. The reduction in speed can be significant. For example, in a piece of glass with $n = 1.5$, $v = c/n = 2.0 \times 10^8$ m/s.

What Is Total Internal Reflection, and How Is It Used?

Think again about looking at the fish underwater. If you try looking at too steep an angle, you can't see the fish at all. In Snell's law, the point at which you can't see the fish any more corresponds to an angle of 90° on the air side, for which Snell's law gives about 49° on the water side. A light ray striking the water's surface (from below) at an angle larger than 49° from the normal suffers **total internal reflection** and doesn't emerge from the water. Total internal reflection occurs at large angles whenever a light ray in a medium of higher index of refraction reaches a boundary with a medium having a lower index.

An important application of total internal reflection is in **fiber optics.** A transparent cable can transmit light signals over great distances with no loss of signal because, each time the light ray reaches the cable's side, it's reflected back inside (Figure 5.5). You can even bend the cable a bit, as long as the bend isn't so large as to let the light ray escape into the surrounding medium, usually air (Figure 5.6). Fiber-optic cables send data from one computer to another by digitizing the light signals—essentially a string of "0" and "1" data bits that are decoded at the destination. Fiber-optic cables are also used to help see into small places—for example, so that surgeons can look into organs or joints to make repairs without having to cut them wide open.

What Is Dispersion?

In many materials (including glass and water), the speed of light varies slightly with the light's color (or wavelength), with shorter wavelengths generally having slower speeds for visible light. Equation (5.2) tells us that the index of refraction must also be wavelength-dependent, with shorter wavelengths having slightly higher n values. Then, by Snell's law, different colors are refracted at different angles, resulting in a separation of colors known as **dispersion.** You've probably seen the spectrum of colors produced by dispersion of white light in a prism. This works because light that you see as white is a combination of the colors of the spectrum, from violet to red. The prism separates them well enough for you to see the individual colors.

Figure 5.5 A light ray sent into a fiber-optic cable experiences total internal reflection at each encounter with the surface, so the ray emerges from the other end.

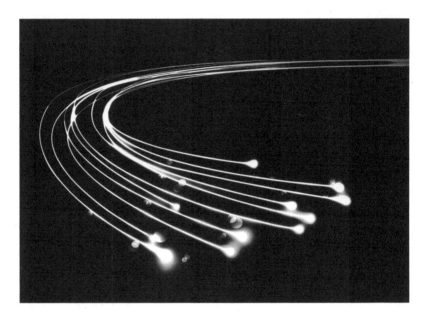

Figure 5.6   A fiber-optic cable carrying light.

Why Do Diamonds Sparkle?

Diamond has an extremely high index of refraction: about 2.4. This makes it especially susceptible to total internal reflection. (Using Snell's law, you can show that internal light is reflected back in unless it makes an angle smaller than 25° with the normal at the diamond's surface.) A ray of light that enters a diamond is likely to bounce around several times before emerging. Thus, light entering different facets of the cut diamond comes out in widely different directions, resulting in the sparkle you see. You may also see bits of color, especially reds and blues. That's due to dispersion when the light ray enters and leaves the stone.

How Do Rainbows Form?

Rainbows come from combining some of the effects we've been talking about—refraction, dispersion, and total internal reflection—in raindrops. Figure 5.7 shows how this works. Raindrops are nearly spherical in shape due to surface tension—not the "teardrop" shape in which they are often depicted. Incoming sunlight is nearly white. It refracts when entering the drop, experiences total internal reflection on the other side, and then refracts again when emerging

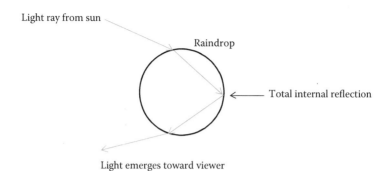

Light ray from sun

Raindrop

Total internal reflection

Light emerges toward viewer

Figure 5.7   Formation of rainbow by total internal reflection and dispersion in a raindrop.

back toward you. (Have you ever noticed that the sun is at your back when you're viewing a rainbow?) The larger refraction of the violet end of the spectrum puts it on the bottom of the rainbow, leaving the red on top. Sometimes you see a fainter secondary rainbow, with the order of colors reversed from the primary bow. The reversal comes from a second internal reflection within the drop.

 HOW DO LENSES WORK?

Lenses work by refracting incoming light rays, either to form images or otherwise redirect the incoming light. In a **converging lens,** one or both surfaces are convex (Figure 5.8), while in a **diverging lens,** one or both surfaces are concave (Figure 5.9). Both lenses shown here have spherically shaped surfaces with the same curvature on both sides, though not all lenses have the same curvature on both sides. The effect of a spherical convex lens is to make incoming rays that are parallel to the lens's symmetry axis bend and converge approximately to a *focal point F* on the other side of the lens, as shown in Figure 5.8. A convex lens's *focal length f,* the distance from the center of the lens to the focal point, decreases as the lens's curvature increases. Thus, a flatter lens has a longer focal length, and a more curved lens has a shorter focal length. Similarly, parallel rays are bent away from a focal point in the diverging lens (Figure 5.9).

How does the lens bend light this way? A glass or plastic lens typically has an index of refraction between 1.5 and 1.8. A light ray in air (with index of refraction 1.0) enters the lens and is refracted, according to Snell's law. Upon leaving the lens, the light ray is bent a second time as it goes from a higher to lower index of refraction. The net effect of both refractions was shown in Figures 5.8 and 5.9.

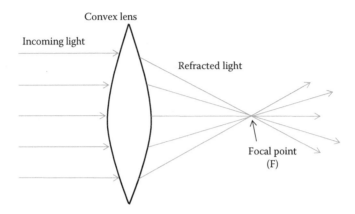

Figure 5.8 Convex lenses make parallel light rays from one side converge to a focal point on the other.

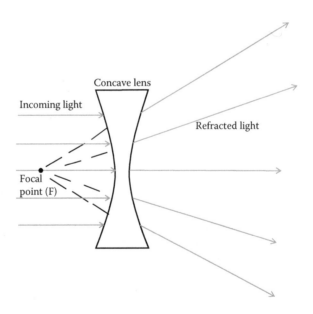

Figure 5.9 Concave lenses make parallel light rays diverge.

Parallel rays that enter farther from the lens's center enter at a larger incident angle; thus, by Snell's law, the outer rays are refracted more, which allows the lens to have a common focal point. The amount of refraction depends on the lens's curvature. A lens with more curvature causes more refraction, so it has a shorter focal length. A flatter lens has a longer focal length.

How Do Converging Lenses Form Images?

Converging lenses are used to form images. There are two distinct cases with vastly different results. When the object (the upright arrow) is placed outside the lens's focal point (Figure 5.10), the image formed is a real image because the incoming light rays converge in one place to form an image that could be viewed on a blank screen placed at that point. The real image is inverted. You can make images of different sizes by moving the object. When the object is far away, the resulting image is small. Move the object closer to the focal point, and the image grows in size.

The situation is very different when a converging lens forms an image of an object placed inside its focal point (Figure 5.11). Now the refractive properties of the lens result in light rays that diverge on the opposite side of the lens, so no real image can be formed. However, if you view the object from the other side, you'll

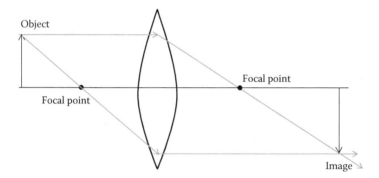

Figure 5.10   Formation of a real image by a converging lens.

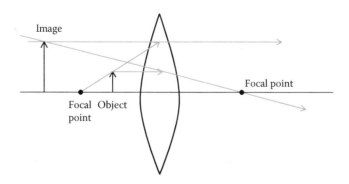

Figure 5.11   Formation of a virtual image by a converging lens.

see a virtual image, where the refracted rays traced backward *appear* to converge. The virtual image formed by a converging lens is always enlarged relative to the object being viewed, and this is what you do when you use a magnifying glass. Fortunately, the image is not inverted in this case, so you simply see it as larger and right-side up. Again, the image size depends on where you place the lens. For the best magnification, put the object just inside the lens's focal point.

Does Your Eye Form Images This Way?

Yes! With convex shapes, your eye's cornea (front surface) and lens focus incoming light rays onto the retina, at the back of your eye, where a real image is formed. Just as shown in Figure 5.10, the real image is inverted. Your brain then processes the images on the retina—including inverting them, so that you don't see things upside down!

What Are Aberration Effects in Lenses?

There are two principal kinds of aberration—lack of perfect focusing—that you encounter with spherical lenses. **Spherical aberration** is an unavoidable effect of the spherical shape of the lens. We noted earlier that the perfect focusing illustrated in Figure 5.8 was an approximation. A careful analysis using Snell's law to find the amount of refraction at each surface reveals that incoming light rays that enter the lens at larger angles of incidence will miss the focal point by increasing amounts. Because spherical aberration depends on the angle of incidence, it is more pronounced in lenses of greater curvature (shorter focal length) or when large-diameter lenses are used.

Chromatic aberration is due to the dispersion of light that contains different wavelengths. Light of different wavelengths has different indexes of refraction, so different colors don't have the same focal length in a given lens. In effect, the lens is acting a bit like a prism, dispersing light of different colors. The amount of chromatic aberration depends on the dispersive properties of the material used in each lens and on the color spectrum of the light being focused. Some modern plastics are designed to have low dispersion, and these are favored in some applications such as eyeglass lenses.

WHAT ARE SOME COMMON REFRACTIVE VISION DISORDERS, AND HOW ARE THEY CORRECTED WITH LENSES?

Some lucky people have great vision for most of their lives, but many of us have imperfect optical systems in our eyes. There are several things that can go wrong that require correction with eyeglasses or contact lenses.

People with **myopia** (nearsightedness) have eyes with a focal length that is too short, so the eye forms real images in front of the retina instead of directly on it. There is something like an image on the retina, but it's blurry, so your vision is blurry, and it gets worse for more distant objects. To compensate for myopia, people wear eyeglasses or contact lenses with diverging lenses. The lenses make incoming light rays diverge slightly before reaching the eye, and the misshapen eye then refracts the diverging rays so that they form clear images on the retina, thus restoring good vision. Conversely, **hyperopia** (far-sightedness) occurs when the eye's focal length is too long. In this case the correction comes from wearing a converging lens, which gives the eye additional focusing power to make a clear image on the retina.

Your eyes do a remarkably good job of adjusting their focusing. The muscles in your eye adjust its optical properties so that nearby and distant objects can be seen clearly. As people get older, the eye's lens becomes less elastic, reducing the range of focal lengths it can access. This condition is known as **presbyopia.** The greatest loss of range tends to be at shorter distances because the eye can't shorten its focal length sufficiently. Accordingly, converging lenses are prescribed for reading and working at shorter distances. If a person is myopic to begin with and develops presbyopia, this calls for bifocals—different prescriptions for viewing longer and shorter distances. That usually means a diverging lens for distance and a slightly less diverging one (that is, diverging with a longer focal length) for reading.

Another common disorder is **astigmatism,** in which the lens is not spherical. The amount of astigmatism is measured by determining the eye's curvature in different planes, ranging from horizontal to vertical. These should be the same, but they vary in a person with astigmatism. The treatment is an eyeglass or contact lens with the opposite asymmetry, to cancel the asymmetry in your eye.

Most people with refractive disorders can have their vision improved dramatically by **laser surgery,** usually the LASIK (laser-assisted in situ keratomileusis) method. If you are nearsighted, the laser is used to flatten the central part of your cornea. For nearsightedness, the laser makes the outer parts of the cornea steeper to increase the curvature. Correcting astigmatism is trickier because it requires careful mapping of the entire cornea so that the laser can restore a spherical shape.

What Does Your Corrective Lens Prescription Mean?

If you have some refractive disorder, say myopia, your optometrist gives you a lens prescription for diverging lenses of the correct shape. Remember that a lens's focal length is related to its curvature, with smaller focal lengths corresponding to more curvature. The lens's **refractive power** is related to its focal length by $P_r = 1/f$, where f is the focal length. The unit for refractive power is the diopter, with 1 diopter = 1 m⁻¹. For example, a converging lens with a focal length of 25 cm (= 0.25 m) has a refractive power $P_r = 1/f = 1/(0.25\ m) = 4.0\ m^{-1}$ = 4.0 diopters.

By convention, a diverging lens is given a negative focal length, and thus it has a negative refractive power (e.g., −4.0 diopters for a diverging lens with a focal length of −25 cm). An eyeglass prescription of −4.0 diopters represents a significantly nearsighted condition. Similarly, a farsighted person gets a prescription with a positive refractive power, such as +2.5 diopters.

If you have astigmatism, you get three numbers in your prescription. The first is the number we've already described, a negative number for nearsightedness or a positive one for farsightedness, in diopters. The second number, called "cylinder," measures the amount of astigmatism, also in diopters. The third number, "axis," gives the orientation of the astigmatic disorder as an angle between 0° and 180°, with 90° oriented vertically.

 ## HOW DO MICROSCOPES WORK?

You've probably used a single convex lens as a magnifying glass. Even the best lenses are limited as magnifiers by aberration effects. You can greatly improve magnification by using a pair of lenses to make a **compound microscope,** shown schematically in Figure 5.12.

In a compound microscope, the objective lens with focal point F_o forms a real image of the object you want to view ("Image 1" in Figure 5.12). Then you place the second lens (the eyepiece, with focal point F_e) so that its focal point

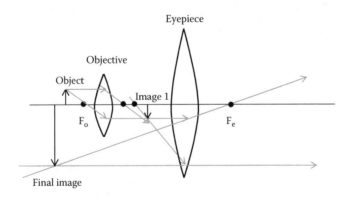

Figure 5.12   Schematic representation of an optical microscope.

is just past the real image, resulting in a much larger virtual image of the real image. Because the real image from the objective was real, it's also inverted. The virtual image in the eyepiece doesn't change that orientation, so the image you see is inverted with respect to the object being viewed. It's also greatly magnified, due to the combined effects of the two lenses. Depending on the quality of the optics, the net magnification can be anywhere from several times up to about 1,500 times, with the best resolution on the order of several hundred nanometers—comparable to the wavelength of visible light. That's good enough to resolve many single living cells and the detail in small integrated circuits. If you have ultraviolet light, you can do even better; its shorter wavelength results in better resolution, provided you have an ultraviolet camera to see the image.

What Is an Electron Microscope?

If ultraviolet light's shorter wavelength allows for better resolution in a microscope, why not make the wavelength even shorter? That's the idea behind an **electron microscope.** Quantum mechanics (Chapter 4) tells us that particles traveling with momentum p have a wavelength $\lambda = h/p$, where h is Planck's constant (6.6×10^{-34} J·s). With their small mass, electrons are perfect for generating the proper wavelengths—shorter than visible light but not so short that they can't be focused. In an electron microscope, electrons can be generated to produce wavelengths of 1 nm or shorter, so they replace light as the imaging tool.

There are two main types of electron microscopes. In a **transmission electron microscope** (TEM), the electron beam passes through a sample, and the detected beam is analyzed to reveal the sample's structure. In a **scanning electron microscope** (SEM), the electron beam is scanned across a portion of the sample's surface. The beam's energy is then converted to a number of different forms that can be detected, including emission of light, x-rays, secondary electrons, or heat; these are then analyzed to form detailed images like the one in Figure 5.13.

 ## HOW DO TELESCOPES WORK?

The basic principle of the telescope is the same as that of a microscope, except that the telescope's objective gathers light to form an image of a distant object instead of a nearby one. Just as in the microscope, the telescope's eyepiece then enlarges the image for viewing. There are two primary types of telescopes. A **refracting telescope** uses a lens as an objective, and a **reflecting telescope** uses a concave mirror for that purpose.

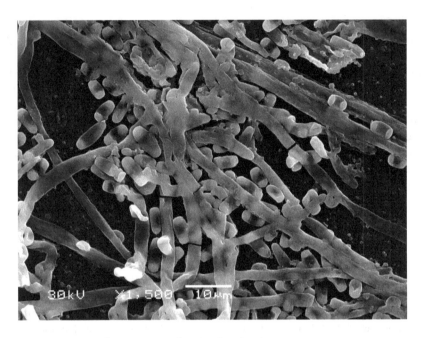

Figure 5.13   Image from an SEM, showing fungi and spores.

How Does a Refracting Telescope Work?

Figure 5.14 shows how a refracting telescope works. Both lenses shown are convex (converging) lenses. The objective lens forms a real image of the distant object you're viewing, and then the eyepiece is used to make a magnified virtual image of the real image. Similar to the microscope setup, you want to place the eyepiece so that its focal point is just past the real image from the objective. This will produce maximum magnification. It's also best to use two lenses with vastly different focal lengths because the telescope's magnification is approximately $-f_o/f_e$, where f_o and f_e are the objective and eyepiece focal lengths, respectively. For example, a pair of lenses with $f_o = 30$ cm and $f_e = 5$ cm has magnification -6, where the minus sign indicates an inverted image. The inversion is because the real image is inverted, but the virtual image in the eyepiece keeps the same orientation.

The telescope shown in Figure 5.14 is a **Keplerian telescope.** It's the simplest kind of telescope to make because all you need are two lenses. But it has the drawback that it produces inverted images. There are other variations of refracting telescopes, such as the Galilean telescope that uses a concave eyepiece to turn the image upright. A sea captain would want to use this kind of telescope to see ships and shore right-side up. You can also use a pair of

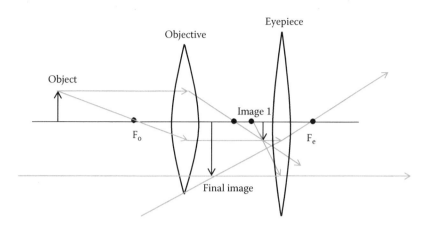

Figure 5.14 Schematic of a refracting telescope with two converging lenses.

reflectors between the two lenses to invert the image, as is commonly done in binoculars.

If you want to do serious astronomical work—looking for fine detail in very distant objects—you need to increase the telescope's light-gathering power by using lenses with large diameters. This brings into play some problems that you don't notice with smaller refracting telescopes. The larger you make the lenses, the more likely it is that you'll notice spherical and chromatic aberrations, which diminish the image's quality. The largest refracting telescopes ever made used objective lenses of over 1 m in diameter. At that size, the glass becomes difficult to support and control, making larger refracting telescopes impractical. These limitations of refracting telescopes explain why reflecting telescopes are used for research in astronomy.

How Does a Reflecting Telescope Work?

A reflecting telescope, also called a Newtonian telescope, is shown schematically in Figure 5.15. Now a concave mirror is used to gather light and form a real image, which can be enlarged by an eyepiece. Note that the mirror is parabolic, not spherical, to eliminate spherical aberration. The fact that it's a reflector eliminates chromatic aberration, too, because light of all wavelengths obeys the law of reflection at the mirror surface. The design shown in Figure 5.15 is typical in that a flat mirror is placed between the concave mirror and the eyepiece. That diverts the image to the side, where it's more easily viewed. This design might be familiar to you if you've dropped in on a public stargazing session in your neighborhood.

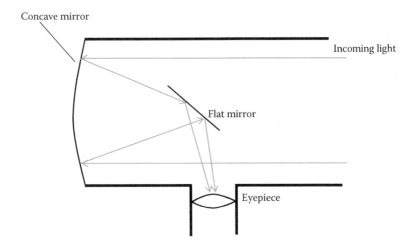

Figure 5.15 Schematic diagram of a reflecting telescope.

Figure 5.16 Radio telescopes in the very large array in New Mexico.

What Are Some Novel Designs for Modern Reflecting Telescopes?

Astronomers started building very large reflecting telescopes for research in the early twentieth century. Edwin Hubble used 100- and 200-inch (referring to the diameter of the reflecting mirror) telescopes to do his pioneering work on distant galaxies, and that work led to our present understanding of the origins and expansion of our universe. You don't have to worry as much about a reflecting mirror's weight because it sits at the bottom of the telescope, compared with a refracting telescope's objective lens, which sits at the top. Even so, it's difficult to keep a very large mirror from sagging under its own weight. Any deviations from a perfect parabolic shape in the reflecting surface produce imperfections in the telescope's images. For that reason some modern telescopes use a segmented primary mirror made of many smaller pieces that are easier to manage. The two Keck telescopes in Hawaii use segmented mirrors that contribute to an overall reflecting diameter of 10 m, with the segments computer-adjusted to maintain the proper curve as atmospheric conditions change.

Can You Use Telescopes to See Other Kinds of (Invisible) Radiation?

Our visible spectrum represents just a small part of the entire electromagnetic spectrum. Stars and other large objects in the universe (including galaxies, quasars, and pulsars) emit a lot of nonvisible radiation, and astronomers have built instruments to see much of this invisible spectrum. In particular, **radio astronomy** has made significant contributions for many generations, most notably the discovery of a weak microwave signal from the edges of the universe with a wavelength of about 1 mm. This is the cosmic background radiation, which provides strong evidence of the Big Bang that occurred 13.8 billion years ago. Radio telescopes need to be large (Figure 5.16) in order to capture radio waves that can be over a meter in length.

Today astronomers build special telescopes dedicated to other parts of the electromagnetic spectrum, including gamma rays, x-rays, ultraviolet, and infrared. Data from each part of the spectrum provide unique information. Many of these telescopes are in space. You've probably seen some of the spectacular images from the Hubble Space Telescope, which began operating in 1990. But you may not realize that there are many other telescopes in space that look at different parts of the electromagnetic spectrum. Placing telescopes above the atmosphere allows them to detect radiation that our atmosphere reflects or absorbs, making it invisible to Earth-based telescopes.

HOW DOES INTERFERENCE REVEAL LIGHT'S WAVE PROPERTIES?

Waves of any kind can interfere with one another. The effects are obvious when you see two boats' wakes passing through one another. You can hear the effects of sound wave interference, when two sounds of nearly the same frequency produce "beats"—a regular increase and decrease in the sound's volume. Light interference isn't as obvious as it is in water and sound waves. You can shine two flashlight beams right through each other, and they seem to pass with no interference. Shine them so that they land together on the same target, and all you see is white—no interference. However, there are a few places where you can notice interference, such as in the colors you see in a soap bubble or oily puddle.

Before we explain those observations, let's think about interference in a more generic way. Light is an electromagnetic wave that varies in space like a sine curve (shown in Figure 2.13 in Chapter 2). When two waves interfere with one another, there are two distinct ways that can happen. If the wave crests overlap with each other, the waves add together in a situation called constructive interference, and you get a wave of greater intensity. But if one wave's crest meets the other's trough, the result is destructive interference—effectively a cancellation of the two waves. For light waves, constructive interference produces visible light, and destructive interference results in a dark area.

To make light waves interfere, you need two sources that send out light waves that eventually meet. It's easiest to see interference if the light sources are **monochromatic,** which means that they produce light of a single wavelength. One reason you don't see those flashlight beams interfere is that white light contains all visible wavelengths, which overlap and combine randomly. A laser is an excellent source of monochromatic light, and a good way to see interference is to shine a laser through a pair of narrow, closely spaced slits. The two slits act like independent sources, and when light waves from the two sources meet on a blank screen on the other side, you see an interference pattern like the one in Figure 4.5 (Chapter 4). Wherever two wave crests arrive together, you see a bright band due to constructive interference, and when a crest and trough meet, the waves interfere destructively, leaving a dark band. This distinctive interference pattern is a convincing illustration of light's wave properties.

Can You See Thin-Film Interference from White Light?

You see interference from white light in a number of places. One place is in the colors that appear in a puddle of water, when the water has a little oil in it. Oil is lighter than water, so it floats on top and spreads out into a very thin layer,

GOING DEEPER—TWO-SLIT INTERFERENCE
AND DIFFRACTION GRATINGS

Analyzing the two-slit interference pattern in more detail will help you see how scientists use interference to study light. Figure 5.17 shows how light from the lower slit has to travel farther than light from the upper slit in order for the two waves to meet at P on the screen. To make P be a point of constructive interference, that extra distance must be a whole number of light wavelengths, or $n\lambda$, where n is an integer. From the small right triangle adjacent to the slits, the extra distance traveled by the lower wave is $d \sin \theta$, where d is the separation between the two slits. Equating the two expressions for distance:

$$n\lambda = d \sin \theta \qquad (5.3)$$

Equation (5.3) makes an important connection between the light's wavelength, the slit spacing, and the "spread" of the interference pattern, which shows up in the angle θ. Changing either d or λ has an effect on θ. For example, for a given pair of slits, you get a broader interference pattern with long-wavelength red light than with shorter wavelength green light. For a given light source, you can broaden the pattern by making the spacing d smaller.

A **diffraction grating** is made by placing a number of parallel lines on a transparent opening, effectively harnessing together multiple two-slit gratings. The combined effect is shown in Figure 5.18, with the result being a sharper and cleaner interference pattern than a single pair of slits can generate. The same relationship (Equation 5.3) governs the diffraction grating's interference pattern. Scientists use diffraction gratings to measure wavelengths emitted by everything from distant stars and galaxies to lasers and fluorescent lights.

perhaps not much thicker than light's wavelength. Natural (white) light coming from above goes through the oil, with some reflecting from the oil–water boundary. This reflected wave interferes with light reflected directly from the top of the oil. The two reflected waves can interfere constructively. The wavelength at which this occurs depends on the oil layer's depth, which is why you see different colors at different places in the puddle. A similar effect allows you to see different colors in a thin soap bubble, where you see interference of light reflected from the inner and outer surfaces of the bubble.

Constructive interference is responsible for the **iridescence** you see in brightly colored butterfly wings and bird feathers. Just as with the oil slick and

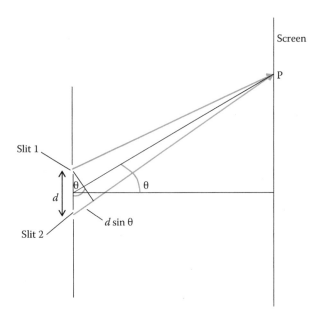

Figure 5.17   Geometry of two-slit interference.

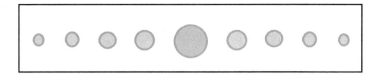

Figure 5.18   Interference pattern from a laser fired through a diffraction grating.

soap bubble, the color depends on interference from the top and bottom layers of a thin structure in that part of the animal.

Are There Any Useful Applications of Interference?

Other than purely scientific applications, interference is used in ways you might not be aware of. Lenses used in eyeglasses, cameras, telescopes, and other applications use an **antireflective coating.** This thin coating—about one-fourth of a wavelength of light—is applied to the lens's surface. Some light that enters the coating reflects from the lens's surface, where it can emerge into the air and interfere with light that reflected directly from the outer coating. The thickness of the coating causes destructive interference between the two exiting waves, which greatly reduces the net reflection you see.

Thin-film interference is used to make some of the "water marks" you see on checks, currency, and legal documents. When viewed from different angles, these marks appear to change color and brightness. The precision of the optical coatings makes it difficult to copy or counterfeit the documents.

WHAT IS DIFFRACTION?

Diffraction is the bending of a wave as it passes an obstacle or through an opening. (Note that this isn't the same as refraction, which is the bending that happens when light goes from one medium to another.) The two-slit interference pattern and diffraction grating work only because light that passes through the openings bends as it does so. The spreading of the light waves allows the interference pattern to form.

The uses of diffraction aren't limited to light from the visible spectrum. Scientists use **x-ray diffraction** to study solids, particularly those that have a regular crystal structure. X-rays are ideal for this purpose because their wavelengths (typically 0.1 to 1 nm) are about the same size as the spacing between layers of atoms in a crystal. X-rays reflected from adjacent planes of atoms interfere with one another, and the resulting interference pattern reveals the orientations of the planes and the spacing between atoms.

WHAT IS POLARIZATION, AND WHY IS IT USEFUL?

If you've ever worn sunglasses, you've most likely used a light polarizer. The plastic in most sunglass lenses contains organic polymers that are oriented linearly. That structure makes them allow through light waves with a preferred orientation, which is what we mean by **polarized light.** Most natural light waves—directly from the sun or reflected from objects on Earth—have no preferred orientation, so we say they are nonpolarized. Most sunglasses have a particular orientation that allows more vertically oriented waves to pass while blocking the horizontal ones. That's because horizontally oriented structures like streets and bodies of water tend to reflect light with a horizontal orientation. Your sunglasses then block that "glare" while reducing the overall amount of light that reaches your eyes.

You can convince yourself of the effectiveness of polarizers if you have two of them that you can hold next to one another. If you hold them so that their polarizers are oriented at 90° to one another, no light will pass because you've effectively blocked waves of all orientations. Now if you rotate the polarizers from 90° to 0°, you'll see the transmitted light increase steadily. These "crossed" polarizers are used in liquid crystal displays (LCDs) found in cell phones, televisions, and other electronic devices. By turning the crossed polarizers on and off electronically, the pixel display goes from dark to light.

WHY IS THE SKY BLUE?

The blue sky is due to **Rayleigh scattering,** a type of dispersion where light scatters from small particles in Earth's atmosphere. Rayleigh scattering varies with the light's wavelength λ as $1/\lambda^4$, so there's a strong preference toward scattering shorter wavelength light. Thus, if you look away from the sun, you'll see blue, resulting from the mix of shorter wavelengths that reaches your eye. When you look closer to the sun, and especially at sunrise and sunset, you see the long-wavelength orange to red colors because the shorter waves have been scattered away from your field of view.

HOW DO LASERS WORK, AND WHAT MAKES THEM DIFFERENT FROM OTHER LIGHT SOURCES?

You see lasers all the time in the barcode scanners in retail stores. You have probably heard how lasers are used extensively in the medical field, especially in eye surgery, and increasingly in cosmetic surgery to remove blemishes or tattoos. There are many other ways you use lasers but don't see them—for example, in electronic devices including CD and DVD players and laser printers. Lasers are used extensively in industry as cutting and imaging tools and in communication via optical fiber networks. There are simply too many other applications to list here.

In a standard design for a laser, light bounces back and forth between two mirrors, one fully reflective and the other only partially reflective, allowing some light to escape—that's the laser's output. The "lasing" material inside the laser and between the mirrors is chosen for its electronic energy levels. Just as with fluorescent lights, transitions between a pair of energy levels produces light having a specific wavelength. In a laser, however, a photon of light bouncing between the mirrors generates **stimulated emission** in the atoms or molecules in the cavity, producing a photon having the same direction and wave phase as the photon that generated it. Because of its common phase, the laser output is said to be **coherent.**

Even though the lasing material may have many energy levels, stimulated emission is restricted to one pair of levels. As a result laser light is highly monochromatic. Incandescent lights produce a full continuous spectrum, and fluorescent lights give off a combination of several different visible wavelengths. Being both coherent and monochromatic distinguishes laser light from other light sources and aids any application that depends on having a precise wavelength or phase information, such as optical interferometry or holography.

How Do CD and DVD Players Work?

If you look closely at a CD or DVD, you can see tiny circular grooves. (You also see a lot of colors, an example of white-light interference because the groove sizes are comparable to light's wavelength!) An even closer examination reveals that the circular grooves are actually small pits in the disk that encode information digitally. Your CD or DVD player has a laser that scans across the rapidly rotating disk, and the information from the reflected signal is converted to sound and/or picture.

The DVD has significantly smaller pit sizes compared with the CD. As you'd expect, that's because it has to carry a lot more information, including both audio and video, on a disk of about the same size. The laser wavelength needs to be shorter to read finer detail. CD players generally use a 780-nm infrared laser, and DVD players use a red 640-nm laser. Newer Blu-ray DVD players use a short violet wavelength of 405 nm.

WHAT IS A LIGHT-EMITTING DIODE?

A light-emitting diode (LED) is a semiconductor device that exploits the diode's **band gap,** a small 1–3 eV difference in energy between the diode's valence band and conduction band. The process is analogous to a fluorescent light, where electrical energy promotes electrons in a gas to excited states, and light is emitted when the electrons fall back to lower energy levels. In the LED, electrical energy moves electrons from the valence to conduction band, and the electron's spontaneous return to the valence band releases a photon.

For many years LEDs have been used in electronic displays, such as calculators. Today they are used increasingly in displays and signs. As a light source, the LED is more energy efficient than traditional forms of lighting. LEDs' brightness and durability give them distinct advantages over other lighting sources, even if their initial cost is higher. Many new autos, trucks, and buses have red LED taillights. Those taillights should last longer than your new car's engine will, and the additional brightness is seen as a safety improvement, compared with old car brake lights that use a red plastic cover over a white light bulb. Older street lights are being replaced by LED lights for the same reason.

For general household lighting, compact fluorescent lights (CFL) are replacing incandescent lights and older fluorescent lights. However, many people dislike the look (especially the color) of light from a CFL. LED light may do better, but this requires some attention to design. Each LED produces light of a single color, like the red taillight. That wouldn't do for household lighting. However, a light that contains multiple LEDs—for example red, blue, and green—produces white light that simulates an incandescent bulb. These will cost more than other lights but will last perhaps 50,000 hours, compared with 8,000 hours for a CFL and 1,000 hours for an incandescent bulb. Further, the LED can produce the same amount of light for less energy input than the CFL.

 ## WHAT IS A SOLAR (PHOTOVOLTAIC) CELL?

A solar cell (also photovoltaic or PV cell) can be thought of as an LED in reverse. In a solar cell, incoming light excites an electron across the semiconductor's band gap to produce electrical energy. Solar cells are attractive as a source of "clean" energy because they emit no greenhouse gases. However, they currently account for less than 2% of electrical energy used in the United States and only about 0.1% worldwide. The semiconductor materials are expensive to make and have a finite lifetime, typically 20–30 years. It's difficult to make solar cells with high efficiency (electrical output relative to solar energy input) because a semiconductor with a specific band gap can use only a small part of the sun's broad energy spectrum. Efficiency is improved by making a layered solar cell, with different layers having different band gaps. However, this adds to the manufacturing cost. Generally solar cells can't compete economically with fossil-fuel energy sources for large-scale energy production, but there is hope that the new semiconductor technology will make solar cells more competitive.

FURTHER READINGS

Hecht, Eugene 2002. *Optics*, 4th ed. San Francisco, CA: Addison Wesley.
Park, David 1997. *The Fire within the Eye*. Princeton, NJ: Princeton University Press.
Pendrotti, Leno S., and Pendrotti, S. J. L. 1998. *Optics and Vision*. San Francisco, CA: Addison Wesley.

Chapter 6

Thermodynamics

When you think of thermodynamics, you probably think of systems that regulate temperature. Your furnace keeps you warm in winter, and air conditioning maintains a comfortable environment in summer. Refrigerating and freezing food preserves it until you're ready to use it. Then you cook it, often at a high temperature. Look around the modern world and you'll see many other thermodynamic processes, from your car's engine burning fuel and then expelling waste heat to power plants that burn fuel to generate electricity. **Thermodynamics** can be defined as the science of heat and temperature and their relation to energy and work. The applications we've mentioned all involve some use or transfer of energy. But leaving out all the modern machines and conveniences, thermodynamics is fundamental to life. You know that your life depends on maintaining a fairly constant body temperature in all environments, but on a molecular and cellular level, basic metabolic processes involve vital thermodynamic reactions.

WHAT IS TEMPERATURE?

Most people have an intuitive sense of **temperature.** Step outside, and you can easily tell the difference between a warm day and a cold day. However, defining the temperature of a particular object or system is not always simple. Temperature is related to **thermal energy,** which is the kinetic and potential energy associated with individual atoms and molecules. (Note that this is distinct from the kinetic and potential energy of a bulk object, such as a ball rolling down a hill.) Generally, temperature goes up as the average thermal energy per atom or molecule goes up. The relationship between temperature and thermal energy can be complicated for some materials, which makes it difficult to devise a general definition for temperature.

It's often easier to think of temperature by comparing two objects in contact with one another because then energy will flow from the higher temperature object to the lower temperature one. The energy that flows spontaneously this way is called **heat.** This example illustrates the close relationship between temperature and energy in thermodynamics.

What Are Some Common Temperature Scales?

Around the world most people measure temperature in either **Celsius** (°C) or **Fahrenheit** (°F). At atmospheric pressure, water's freezing point is close to 0°C (32°F), and its boiling point is close to 100°C (212°F). Accordingly, a temperature difference of 1°C is equal to 9/5 of 1°F.

However, knowing just the freezing and boiling points doesn't define the entire temperature scale. You need to be able to measure temperatures in between and outside that range. One type of material for which it's straightforward to measure a wide range of temperatures is an **ideal gas,** which is defined as a gas with sufficiently low density and high temperature that the molecules making up the gas interact infrequently with one another. Although those conditions might sound constraining, most common gases (for example, nitrogen, oxygen, and helium) exhibit essentially ideal behavior at atmospheric pressure and temperature.

Ideal gases follow the **ideal gas law,** which relates the gas's pressure P, volume V, amount of gas measured in moles n, and temperature T:

$$PV = nRT \qquad (6.1)$$

where $R = 8.314$ J/(K·mol) is the ideal gas constant.

If you measure a gas's pressure as a function of temperature at constant volume, the result will be as shown in Figure 6.1. The function is linear, as predicted by the ideal gas law. As the gas's temperature is lowered, it will liquefy at some point, but extrapolating the graph beyond that point allows you to find **absolute zero** at –273.15°C, regardless of the type of gas used in the experiment. The **kelvin** (K) temperature scale measures absolute temperatures, meaning that 0 K falls at the absolute zero point. Therefore, all temperatures are positive on the kelvin scale. The size of the kelvin degree is the same as the Celsius degree, so 0°C = 273.15 K.

Notice that the ideal gas law only makes sense if absolute temperatures are used because it doesn't contain an additive constant to adjust for absolute zero and doesn't allow for negative temperatures. K is the SI unit for temperature and is used in most scientific applications. Although it's named for a person (William Thomson, Baron Kelvin of Largs), the unit is spelled out in lower case kelvin, just like the units newton and joule. It's conventional to omit the degree symbol (°) and to write the unit as K, unlike the units °C and °F.

Once you've defined a zero point and the size of a degree for a temperature scale, you can calibrate thermometers to that scale. There are several types in

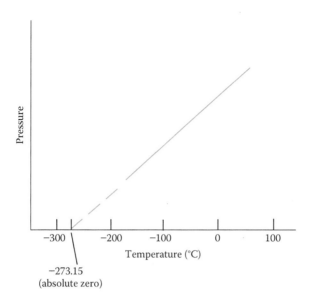

Figure 6.1 Pressure versus temperature for an ideal gas.

common use. A liquid-bulb thermometer uses the property of thermal expansion (see later discussion). Increasing temperature makes the liquid (such as mercury or alcohol) expand into a thin column, where there are markings corresponding to the liquid's temperature. A dial thermometer uses thermal expansion of a metal coil to move a pointer to different temperature markings. Many thermometers today are electronic with digital displays. These contain a material with some electronic property, such as electrical resistance, that varies reliably as a function of temperature. That electrical property is calibrated to the temperature display.

 ## WHAT IS THE KINETIC THEORY OF GASES?

The kinetic theory of gases expands on the empirical ideal gas law and helps explain more of a gas's properties. This theory assumes particles (atoms or molecules) of a single type, each having the same mass m. This is reasonable because all atoms or molecules of a single type are identical to one another (Chapter 4). For example, the common oxygen molecule O_2 has two oxygen atoms, each with a mass of about 16.0 u (atomic mass units), so the molecule's mass is 32.0 u. One mole of O_2 has a mass of 32.0 g.

Kinetic theory is valid for a sample containing a large number of particles. That's an easy criterion to satisfy. One mole is by definition Avogadro's number $N_A = 6.02 \times 10^{23}$ particles, so even a small fraction of a mole has a very large number of particles. The particles move randomly within the container in such a way that the gas's density is fairly constant throughout. Collisions between gas particles are infrequent, and when they do occur, they're elastic. Collisions between particles and the container walls are also elastic. That's the source of the pressure P (force per unit area) on the walls because, as a particle strikes the wall and a force from the wall makes the particle rebound, Newton's third law requires an equal force on the wall in the opposite direction (i.e., outward).

Kinetic theory assumes that the gas is ideal, so the distances between particles are much larger than the particle size. Thus, potential energy associated with interparticle forces is negligible, and the gas's thermal energy is due entirely to the kinetic energies of the particles. At a given temperature T, the gas particles don't all have the same energy, but their *mean* kinetic energy $\overline{K}$ is directly proportional to temperature:

$$\overline{K} = \frac{3}{2} kT \tag{6.2}$$

where $k = 1.38 \times 10^{-23}$ J/K is **Boltzmann's constant.** Individual particles each have the same mass m, but have different speeds v. The mean kinetic energy is related to mass and speed: $\overline{K} = \frac{1}{2} m\overline{v^2}$, which suggests that there is a **root-mean-square (rms) speed** v_{rms} defined as

$$v_{rms} = \sqrt{\overline{v^2}} \tag{6.3}$$

You can think of v_{rms} as a typical particle speed and, by Equation (6.2), it increases with temperature. Nitrogen and oxygen molecules in air have rms speeds of about 500 m/s at room temperature.

Boltzmann's constant k, which relates the energy to the temperature of individual particles, plays the role that the gas constant R plays for bulk gases. In fact, the two differ by only a factor of Avogadro's number: $k = R/N_A$. Because the number of particles N is related to the number of moles n by $N = N_A n$, the ideal gas law can be expressed as

$$PV = NkT \tag{6.4}$$

You can think of Equation (6.4) as the particle or molecular form of the ideal gas law.

GOING DEEPER—MAXWELL SPEED DISTRIBUTION

The rms speed is a typical particle speed in an ideal gas, and higher temperatures generally correlate to faster particle speeds. At any temperature, however, particle speeds are distributed over a wide range that stretches above and below the rms speed. In about 1870, Scottish physicist James Clerk Maxwell used a statistical analysis to derive the **Maxwell speed distribution,** which gives the relative probability that a gas particle with mass m has any particular speed v at temperature T:

$$F(v) = 4\pi \left(\frac{m}{2\pi kT} \right)^{3/2} v^2 e^{-\frac{mv^2}{2kT}} \qquad (6.5)$$

The distribution is a product of v^2, which increases as a function of v, and an exponential function that decreases as a function of v. This product makes the distribution approach zero at very low and very high speeds, as you might expect on physical grounds, with a single peak in between (Figure 6.2). The distribution is asymmetrical, with a broader "tail" at higher speeds. This explains why the rms speed—the mean of the speed **squared**—is to the right of the graph's peak, which represents the most probable speed.

Figure 6.3 shows the Maxwell speed distributions for identical gases at two different temperatures. At the higher temperatures, there are more particles moving at higher speeds, but there's still a significant overlap between the distributions. Notice that the higher temperature distribution is also broader than the lower temperature distribution. Higher temperature correlates to more randomness in the particle speeds.

WHAT IS THERMAL EXPANSION?

Most solids and liquids expand when their temperature increases. Over reasonably small temperature ranges, the expansion in any direction (or in volume) is fairly linear as a function of temperature. A common example of thermal expansion is the liquid in a liquid-bulb thermometer, in which you see a liquid moving along a linear temperature scale marked on the tube. Different materials exhibit different rates of expansion. The **coefficient of volume expansion** measures a material's fractional volume change per degree of temperature change. Generally, liquids expand more than solids over the same temperature range.

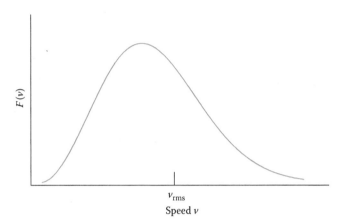

Figure 6.2   Maxwell speed distribution, showing the rms speed.

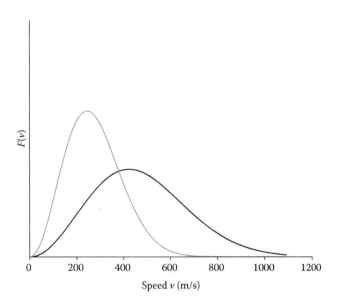

Figure 6.3   Maxwell speed distribution for gases at two different temperatures: 100 K (blue) and 300 K (black). At the higher temperature, more of the distribution is shifted to faster speeds.

Liquid water is an interesting exception. At temperatures of 4°C and above, water behaves normally and expands as its temperature rises. But between 0°C (the freezing point) and 4°C, water has the opposite behavior, expanding as it gets colder. This unusual behavior affects the formation of ice. Near the freezing point, colder water has lower density, which makes ice form at the water's surface rather than below. Once the ice forms, it stays at the surface because its density (about 920 kg/m³) is lower than water's (1000 kg/m³). The surface ice insulates the water below, which restricts the rate of ice formation. This keeps an entire lake from freezing and allows aquatic life to survive below the ice through the winter. When the water is warmer (all above 4°C), the cooler water is denser and hence closer to the bottom, making it cooler as you dive down.

Engineers and construction workers need to take thermal expansion into account or else catastrophic failure can result. Bridges, road surfaces, and railroad tracks have to be built with gaps that allow the materials to expand on hot days. If those gaps are too small, the expanding material can buckle and deform. Larger buildings contain expansion joints that allow the building to flex under thermal stress.

WHAT ARE THE UNITS FOR THERMAL ENERGY, HEAT, AND FOOD ENERGY?

Thermal energy (the kinetic and potential energy associated with individual atoms and molecules) and heat (the energy that flows spontaneously from warmer to cooler objects) are both forms of energy, and therefore the SI unit for both thermal energy and heat is the joule (J). Sometimes the non-SI unit calorie (cal) is used in thermodynamics, with the conversion factor of 1 cal = 4.186 J. Calories are convenient in some applications because 1 cal is the amount of energy needed to raise by 1°C the temperature of 1 g of liquid water.

One calorie is a fairly small amount of energy, so you often see energy expressed in thousands of calories or kcal. That's actually the unit commonly used to express the energy content of food, where the unit kcal may be written Cal, with an uppercase C to distinguish it from the standard calorie (cal). Thus, one "food calorie" = 1 Cal = 1 kcal = 1000 cal.

In the United States, the British thermal unit (btu) and therm are sometimes used, particularly in reference to heating and cooling of interior building spaces. 1 btu = 1055 J, and you are most likely to see a btu rating on appliances used for cooking, refrigeration, or air conditioning. 1 therm = 100,000 btu or 105.5 MJ. Therms are used most often to describe the energy content of natural gas.

WHAT ARE HEAT CAPACITY AND SPECIFIC HEAT?

When an object absorbs heat, it generally experiences a temperature increase. (The exception to this is when a phase change occurs, such as melting ice.) The temperature change ΔT is proportional to the heat absorbed Q, so

$$Q = C\Delta T \qquad (6.6)$$

with C called **heat capacity.** Heat capacity depends on the type and amount of material. It's also temperature dependent, although for small temperature changes the specific heat can be approximated as constant for a given material.

Specific heat c is the heat capacity per unit mass, or $c = C/m$. Because specific heat doesn't depend on mass, it's a characteristic of the type of substance itself. In terms of heat capacity, the relationship between heat absorbed and temperature change is

$$Q = mc\Delta T \qquad (6.7)$$

It takes 1 cal of heat to raise by 1°C the temperature of 1 g of water, so water's specific heat is 1 cal/(g·°C). (You may also see this unit written as cal/(g·K). Measuring specific heat involves changes in temperature, and because the kelvin and degrees Celsius units are the same size, either one may be used in this context.) Using the SI units of joule and kilogram, water's specific heat is 4186 J/(kg·°C).

Solids, especially metals, tend to absorb heat readily. This gives them a much smaller specific heat than that of water, so if you transfer the same amount of heat to the same mass of water and a piece of metal, the metal's temperature increases much more than the water's. Similarly, land on Earth has a much smaller specific heat than water. Large bodies of water (especially the oceans) take a long time to get warmer and cooler with seasonal changes, while land areas far from the ocean suffer greater temperature extremes.

How Is Specific Heat Measured for Gases?

When a gas is heated, its temperature increases, just as in solids and liquids. However, the ideal gas law shows that if a fixed amount (n mol) of a gas's temperature increases, its pressure and volume are also subject to change. To simplify matters, the specific heat of a gas is normally measured with one of those parameters—either pressure or volume—held constant. This leads to two different values of specific heat for a gas: c_V at constant volume and c_p at constant pressure. It's customary to measure the amount of gas in moles rather than a mass in kilograms, so the units for a gas's specific heat (either c_V or c_p) are J/(mol·°C). The result is called **molar specific heat.**

For any particular gas, c_p is larger than c_V. That's because when a gas is heated at fixed volume, all the energy added to the system goes into kinetic energy of the gas molecules. In the constant-pressure case, the ideal gas law requires that the gas's volume increase proportionally to the temperature. Thus, some of the added energy in the system goes into work required to expand the container. Relative to the constant-volume process, more energy is required to raise the temperature by a given amount, and as a result $c_p > c_V$.

What Is the Equipartition Theorem?

Comparing the molar specific heats c_V of different gases reveals a striking pattern. Monatomic gases (such as helium and argon) all have about the same specific heat: 12.5 J/(mol·°C). For diatomic gases (such as nitrogen and oxygen), the specific heat varies within a fairly narrow range of about 20–21 J/(mol·°C).

Monatomic gases can be understood using kinetic theory. Because this type of gas consists of single atoms, you can think of them as small particles bouncing around inside their container, so the thermal energy is just the translational kinetic energy of all the molecules (or atoms). A monatomic gas has kinetic energy $3/2\ kT$ per molecule (Equation 6.2), equivalent to $3/2\ RT$ per mole. Heat added to a monatomic gas at constant volume goes entirely into kinetic energy of the molecules, so the molar heat capacity should be $3/2\ R$. With the gas constant $R = 8.314$ J/(K·mol), then $3/2\ R = 12.5$ J/(mol·°C), which is right on the measured value.

In a diatomic gas, each molecule consists of two atoms, connected by a linear bond. Think of the molecule as two balls held together with a thin rod. Heat added to a diatomic gas also shows up as kinetic energy, but the kinetic energy has two forms: translational motion of the molecules traveling through the container and rotational motion of the molecule (the "rod" spinning). This additional kinetic energy requires more heat added per mole of gas to create a given temperature change and hence a larger specific heat for the diatomic gas.

But why do so many diatomic gases have nearly the same molar specific heat? This is an example of the **equipartition theorem,** which says that molar specific heat is ½ R for each molecular degree of freedom. A degree of freedom is an independent way for a molecule to gain energy. In the monatomic gas, each molecule is free to move in three dimensions. This gives it three degrees of freedom and a molar heat capacity of $3 \times$ ½ $R = 3/2\ R = 12.5$ J/(mol·°C), as measured. A diatomic molecule can rotate about either of the two axes perpendicular to its bond, giving it two extra degrees of freedom for a total of five. Thus, the equipartition theorem predicts that $c_V = 5 \times$ ½ $R = 5/2\ R = 20.8$ J/(mol·°C), which is close to many measured values.

The equipartition theorem also does pretty well for some solids that have a cubic lattice. In the solid state, each atom can vibrate back and forth along the bonds with its neighbors in each of three perpendicular directions. This gives it three degrees of freedom for vibrational motion, but there are an additional three degrees of freedom for the potential energy that changes when the bonds are compressed or extended. With six degrees of freedom, $c_V = 6 \times \frac{1}{2} R = 3R = 24.9$ J/(mol·°C), which is close to the measured value for cubic solids such as copper.

 ## WHAT ARE PHASE CHANGES?

Under ordinary conditions, matter is found in three distinct phases: solids, liquids, and gases. You're used to seeing all three for water: ice, liquid water, and steam. Living fairly close to 1 atm pressure all the time, the way to change the phase of water is to change its temperature. At that pressure, water freezes at 0°C and boils at 100°C. But as the phase diagram in Figure 3.1 in Chapter 3 shows, the boundaries between the phases depend on both pressure and temperature. At pressures less than 1 atm, water boils at a noticeably lower temperature. For example, at an altitude of 1 mile (1.6 km), normal air pressure is about 0.83 atm, and water boils at 95°C. If you put an open container of water inside a closed jar and begin pumping out the air with a mechanical vacuum pump, you'll see water boil at room temperature (20°C) when the air pressure reaches 0.02 atm.

Some features on water's phase diagram are typical of many materials. The phase boundaries meet at the **triple point,** where the three phases coexist. Water's triple point is at 0.01°C and 0.006 atm, a pressure low enough that you don't notice the triple point in everyday life. At the **critical point,** the liquid–gas boundary ends because the two phases have the same density and become indistinguishable. Water's critical point is at 374°C and 220 atm—also far outside your everyday experience.

Water's phase diagram is atypical in one respect: the negative slope of the solid–liquid boundary. That means that the freezing/melting temperature drops as pressure increases, a fact due to water's anomalous thermal expansion as it freezes and the lower density of the solid phase. A more typical phase diagram is the one for CO_2, shown in Figure 3.2 (Chapter 3), where the slope of the solid–liquid boundary is positive. Like water, carbon dioxide has a triple point and a critical point. But notice that there's no liquid phase at pressures anywhere near 1 atm. Thus, at atmospheric pressure, a warming piece of solid CO_2 passes directly from the solid phase into a gas, a process called sublimation. You may have seen this happen. Solid CO_2 is commonly called "dry ice" and is used as a refrigerant. Take a block of it out of the freezer, and you'll see the vapor coming off as it sublimates.

What Is Latent Heat?

Latent heat is the energy associated with crossing a phase boundary. When you melt ice, you have to supply energy in order to break the molecular bonds in the solid. Going in the opposite direction, freezing water requires that you remove the same amount of energy from the liquid. Similarly, turning a liquid into a gas requires energy to overcome the attractions between molecules that keep the substance in its condensed liquid state, and returning from gas to liquid requires removal of energy.

The amount of energy required is different at each phase boundary, so there are two different kinds of latent heat: latent heat of fusion (L_f) for the solid–liquid transition and latent heat of vaporization (L_v) for the liquid–gas transition. In each case, latent heat is defined as the required heat energy per unit mass, so the SI units for both kinds of latent heat are joules per kilogram. For water at 1 atm, $L_f = 3.3 \times 10^5$ J/kg and $L_v = 2.3 \times 10^6$ J/kg. It makes sense that L_v is much higher. Throw some ice into an empty pot on your hot stove. You can melt the ice fairly quickly, but boiling away all the water takes much longer. For water or any substance, the values of latent heat vary as you move along the phase boundaries.

What Is Evaporative Cooling?

Sweating makes you feel cooler on a hot day or during vigorous exercise. It's not the liquid itself that's cooler because it comes out onto your skin at body temperature. However, when the liquid on your skin evaporates, the latent heat associated with the phase change removes thermal energy from your body to the surrounding air. Mechanical refrigerators and air conditioners use evaporation of a working fluid to transfer thermal energy in a similar way.

Evaporation works, even though the liquid is well below the boiling point, because some faster moving molecules can escape the fluid as long as the air (or other surrounding medium) isn't saturated. Evaporation of water from oceans and other bodies of water plays a vital role in Earth's climate. It's promoted by sunlight, which warms the water's surface and increases the rate of evaporation. At higher altitudes, where the air pressure and temperature are lower, water vapor in clouds condenses and falls back to the surface as rain or snow.

 ## WHAT IS THE FIRST LAW OF THERMODYNAMICS?

The first law of thermodynamics deals with changes to a system's **internal energy,** which includes thermal energy (the kinetic and potential energy of individual molecules) and the potential energy associated with interactions between molecules. The **first law of thermodynamics** says that the change

ΔU in a system's internal energy is equal to the sum of the heat Q added to the system and the work W done on the system. Symbolically, the first law is

$$\Delta U = Q + W \qquad (6.8)$$

Heat Q is the energy that flows spontaneously from a warmer to cooler object. Any other energy that affects a system's internal energy is included in the work W. In this context, work includes not only mechanical work—such as a piston compressing a gas in an engine's cylinder—but also any other type of energy transfer that is not heat. For example, cooking your food in a microwave oven is considered to raise the food's internal energy through work (W). Cooking in a conventional oven, in which your food absorbs energy from the hot oven, is counted as heat (Q).

The first law is a simple statement of the wider principle of conservation of energy, applied to thermodynamic processes. It says that a system's internal energy goes up or down by an amount equal to the energy flowing in or out. Both Q and W are positive for energy flowing into the system and negative for energy flowing out.

Internal energy generally correlates to the system's temperature: Increase internal energy and the temperature rises; decrease internal energy and the temperature falls. For simple gases, the connection is straightforward. For example, the only internal energy in a monatomic gas is thermal energy, which, according to the equipartition theorem, is $3/2\ RT$ per mole. Thus, a change in internal energy ΔU due to heat and/or work results in a temperature increase ΔT, with $\Delta U = 3/2\ R\Delta T$ per mole. Similarly, $\Delta U = 5/2\ R\Delta T$ per mole for diatomic gases. For more complex systems, there are no such simple relationships, but you can be confident that putting your food into the oven (microwave or conventional!) will raise its temperature.

What Are Conduction, Convection, and Radiation?

There are three distinct mechanisms for heat flow: conduction, convection, and radiation. Each one involves the spontaneous flow of energy from a warmer to a cooler region.

Conduction is heat flow through direct contact and is the result of energy transfer from faster moving to slower moving molecules. Thus, you observe the direction of heat flow from warmer to cooler objects or from warmer to cooler regions within a single object. For example, heat one end of a metal rod, and conduction through the metal will raise the temperature throughout. Different materials conduct heat at different rates. The rate of heat flow through a material is proportional to temperature difference and to the material's **thermal conductivity,** which measures how effectively the material conducts. Thermal conductivity is high for metals because the free electrons that conduct electricity also help transmit heat. Gases and some low-density materials such as fiberglass have low thermal conductivity. That's why fiberglass is used to insulate walls and attic spaces and a gas-filled gap makes a double-pane window a good thermal insulator.

Convection is heat transfer by the motion of a fluid (liquid or gas). There are many examples of convection in nature and in the machines we use. One important application is in automobiles, where a liquid coolant takes heat from the engine. Many buildings and homes use a gas or electric furnace to heat air, which is then forced through the building with fans. As they are heated, fluids become less dense and therefore tend to rise, which helps create natural convection patterns in the atmosphere and oceans. Respiration involves convection because you normally take in air that's cooler than your body temperature. Then, the gases that you exhale have been warmed by your lungs, resulting in a net outward energy flow.

Radiation is carried by electromagnetic waves. One example is blackbody radiation (Chapter 4), where electromagnetic radiation transfers energy from warmer to cooler objects. This is the principle behind old-fashioned incandescent lights, which use a hot filament to produce light and, in doing so, warm their surroundings. Blackbody radiation is a type of heat flow, but not all electromagnetic waves are characterized as heat. That's because electromagnetic waves are also produced by electronic transitions in atoms (as in fluorescent lights and x-rays), nuclear transitions and reactions (Chapter 7), and oscillating electric currents that make radio waves and microwaves (Chapter 2).

How Does a Thermos Bottle Work?

A thermos bottle insulates its contents from the effects of conduction, convection, and radiation. A vacuum layer between the contents and the outside limits conduction. Convection is prevented by a tight seal. Radiation is restricted by metallic walls, which reflect incoming or outgoing radiation. Scientists use a higher quality version of a thermos bottle called a Dewar flask to store low-temperature "cryogenic" fluids, such as liquid nitrogen, which boils at 77 K at 1 atm. Liquid helium, with a boiling point of 4.2 K, is contained in a special Dewar flask with a surrounding layer filled with liquid nitrogen, which further limits losses by radiation.

What Is the Greenhouse Effect?

A greenhouse is a structure with an outer shell made of glass or other transparent material that allows sunlight to enter to warm the interior. Like any enclosed structure, the walls and roof prevent energy flowing out by convection, but the glass also reflects infrared radiation that strikes it from the inside back into the structure. This keeps the interior warm to promote plant growth.

The **greenhouse effect** describes a similar situation in which the whole Earth is warmed by sunlight, and the transparent atmosphere plays the role of the glass shell. Like the greenhouse's glass, it allows visible light from the sun to reach Earth but absorbs much of the escaping radiation, which is then reemitted in random directions, with some sent back toward Earth. We rely on the greenhouse effect to keep us from getting too cold at night. With no atmosphere, our moon and the planet Mercury are very cold on their unlit halves, with nothing to prevent radiation into space. Fortunately, Earth's atmosphere keeps that from happening here. There's now concern, however, that excessive emission of CO_2 and other "greenhouse gases" by human activity results in sending too much radiation back to Earth, leading to climate change.

 ## WHAT IS THE SECOND LAW OF THERMODYNAMICS?

The **second law of thermodynamics** is more subtle than the first law because it can be stated in several distinct but equivalent forms. Scientists came to understand the second law in the nineteenth century, based in large part on their observations of thermodynamic processes in machinery. One version (called the Kelvin statement of the second law) says that it's impossible in any reversible process to convert any given amount of thermal energy entirely into mechanical energy, or work. Another version (the Clausius statement of the second law) says that no process can take heat from a colder to a warmer object without the expenditure of work. This statement is not violated by the operation of a refrigerator or air conditioner, which removes heat from a cooler space but must do mechanical work and expel waste heat in the process. Both the Kelvin and Clausius statements of the second law are consistent with the observation that the spontaneous flow of heat is from warmer to cooler objects.

Another subtlety of the second law is that it expresses not a certainty, but rather a probability or likelihood that physical processes will move in a certain direction. For macroscopically sized things, the probability that processes will evolve as predicted by the second law is extremely high—so high that it's virtually certain you'll never observe a violation. For example, if you throw a hot chunk of metal into cold water, you can expect that heat will flow from metal to water until the temperature is equal throughout the mixture. The reverse process, with the water transferring heat to the metal to make the water colder and the metal hotter, is so unlikely that you can be confident it won't happen. But for smaller systems, such as those containing just a few molecules, infrequent and temporary violations of the path predicted by the second law have a measurable chance of happening. For that reason, a more complete understanding of the second law requires an understanding of probability and the concept of **entropy.**

What Is Entropy?

Entropy can be defined in two ways. First, in terms of a thermodynamic process, when heat Q flows into a system at temperature T, then the change in entropy ΔS is

$$\Delta S = Q/T \tag{6.9}$$

Note that Equation (6.9) expresses only changes in entropy and doesn't tell you the object's total entropy S. For many systems, S can be difficult to define, but fortunately in thermodynamics it's often sufficient to understand how entropy is changing. The absolute temperature T is always positive, so ΔS is positive when heat flows into a system, and ΔS is negative when heat flows out.

The second law of thermodynamics can be restated in terms of entropy as follows: In any thermodynamic process, the net entropy change for a closed system is

$$\Delta S \geq 0 \tag{6.10}$$

Stated in words, entropy always tends to increase. Thus, while energy is a conserved quantity, entropy is **not** conserved because it increases in time. Think again of what happens when you throw a piece of hot metal into cold water. Heat flows into the water, corresponding to an entropy increase. Heat flows out of the metal, which decreases entropy. The net entropy of the system (metal + water) increases, however, because the positive gain is more than the negative loss. This is consistent with the Clausius statement of the second law because the flow of heat from a warmer to a colder body results in a net entropy gain. The second law gives a preferred direction to changing entropy, and thus it defines what physicists call an **arrow of time** that distinguishes the future from the past.

The second way to define entropy is in terms of probabilities. Think of a gas that consists of just three molecules, shown in Figure 6.4. In a gas, molecules move randomly inside the container. Any one molecule has a 50–50 chance (probability = ½) of being on the left side. The probability that all three will be on the left side is $(1/2)^3 = 1/8$. The probability that all three are on the right is likewise 1/8. The probability that exactly one molecule will be on the left side (or right) is 3/8 because any one of the three could be on the left (or right). To summarize, the probabilities that there will be zero, one, two, or three molecules on the left side are 1/8, 3/8, 3/8, and 1/8, respectively. Notice that there's

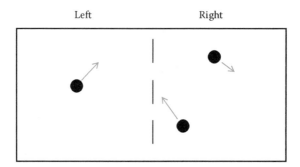

Figure 6.4   A three-molecule gas.

a higher probability of having nearly equal numbers on the two sides. For a macroscopic container of gas that contains a huge number of molecules, it becomes **much** more probable that the two sides contain nearly equal numbers of molecules because there are many more ways to place the molecules in the container so that their populations are close to equal.

It's the number of possible configurations (Ω) of the system that leads to the statistical definition of entropy:

$$S = k\ln\Omega \qquad (6.11)$$

where, as before, k is Boltzmann's constant and the symbol ln is the natural logarithm function. The entropy S grows as the number of configurations increases.

As an example, think of a gas like the one in Figure 6.4 but with a much larger number of molecules and with a fixed partition between the two sides. Suppose you start with all the gas on the right side and the left side empty. If you open the partition so that molecules are free to occupy both left and right sides, you'll soon find that about half are on each side. That's the most likely situation, with the largest number of configurations Ω and the highest entropy S. From the time you open the partition, the system evolves from a low-entropy state (with all the molecules on the right side) to a state of maximum entropy. That's in agreement with the second law, which says that entropy increases in time.

Sometimes entropy is associated with disorder. Thinking again of the gas in Figure 6.4, you might say it's more ordered when all the molecules are on one side and more disordered when they're mixed throughout the container. With this interpretation, the second law's requirement that entropy increase in time means that things become more disordered. Going in the opposite direction to create a more ordered system requires that you generate an equal or greater amount of disorder somewhere else, in order to compensate. The interpretation of entropy as disorder is controversial because disorder can't be quantified and can be subjective. You're on firmer ground if you can define probabilities and use Equation (6.11) to compute entropy.

WHAT IS A HEAT ENGINE?

A **heat engine** is any device that uses heat to produce mechanical work. This includes such common and important devices as (conventional) car engines, train and airplane engines, and electric power plants. For example, a car engine burns gasoline, and the hot gases produced by combustion do work, ultimately transferring energy through the pistons and gears to move the car. In a coal, gas, or nuclear power plant, hot steam does work to turn a turbine, and this mechanical energy is converted to electrical energy by a generator.

The simplest model of a heat engine consists of a hot reservoir at temperature T_h and a cold reservoir at T_c. Energy flows from the hot reservoir to the engine, with part of the energy used to do work and the rest expelled as waste heat into the cold reservoir (Figure 6.5). According to the second law of thermodynamics, no heat engine can be 100% efficient because some of the engine's energy is not available to do work and must be expelled as heat. The heat engine's efficiency depends on many parameters, but in a classic model called the **Carnot cycle,**

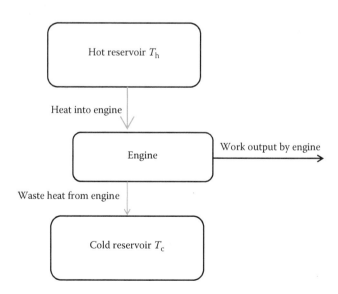

Figure 6.5   Model of how a heat engine works. Some of the energy input is turned into mechanical work, and the rest is expelled as waste.

the engine's maximum efficiency depends in a simple way on the reservoir temperatures T_h and T_c. The Carnot efficiency is

$$e = 1 - \frac{T_c}{T_h} \qquad (6.12)$$

This shows that greater efficiency is achieved for a larger temperature difference. (Note that absolute temperatures are required in Equation 6.12.) Although your car engine doesn't operate on a Carnot cycle, it's still true that it's more efficient after "warming up" than when it has just started.

How Do Refrigerators, Air Conditioners, and Heat Pumps Work?

A **refrigerator** is like a heat engine run in reverse. That is, with an electrical energy input (instead of a work or energy output), heat is removed from a cold reservoir and sent to a hot reservoir. Because energy is used in the process, it does not violate the second law of thermodynamics to take heat from a colder to a warmer place.

In a real refrigerator, the energy input is used to run a mechanical compressor, which compresses a working fluid with a low boiling point. After compression takes the fluid from gas to liquid phase, it can absorb heat from the cooler environment. The fluid is then allowed to expand and return to the gas phase and, in doing so, the heat of vaporization returns thermal energy to the warmer environment. Once this is accomplished, the cycle can repeat with another compression. An **air conditioner** operates using the same principles, except that the cold reservoir is now a room or entire building instead of a closed refrigerator compartment. The air conditioner returns its waste heat to the warmer outside air.

A **heat pump** is similar in principle to the refrigerator and air conditioner because it takes heat from a colder to a warmer place. As a heating device, the heat pump is more efficient than a conventional furnace. That's because a relatively small amount of input energy is needed to move heat in the direction of colder to warmer, whereas a conventional furnace can at best convert just its input energy to heat. In summer, the heat pump can be reversed to operate as an air conditioner, taking heat from the cooler indoors to the warmer outdoors. Thus, you gain efficiency by using the same device year-round.

What Is a Fuel Cell?

Water (H_2O) can be separated into hydrogen and oxygen in the process of **electrolysis,** where electric current supplies the energy needed to break the molecule, and then the gases (H_2 and O_2) are collected at the cathode and anode. The process is reversed in a **fuel cell,** in which hydrogen and oxygen are combined, with electrical energy released and water formed as a by-product. Some cars are now being made with fuel cells replacing gasoline engines. These hydrogen-powered cars exhaust only water and do not emit greenhouse gases. Large-scale use of hydrogen-powered cars will require an infrastructure for producing (by electrolysis or other means), transporting, and delivering hydrogen as widely as gasoline is now distributed.

WHAT ARE QUANTUM STATISTICS?

In the study of atoms and subatomic particles, quantum mechanics (Chapter 4) describes individual particles using wave functions. Generally, it's not possible to resolve wave functions of neighboring particles, and therefore the particles are indistinguishable. To see why this affects particle statistics, consider again the three-molecule gas shown in Figure 6.4. That gas was dilute enough that it was safe to assume that the molecules were distinguishable, and that assumption led to probabilities of 1/8, 3/8, 3/8, and 1/8, respectively, of finding zero, one, two, or three molecules on the left side of the container. Such a computation is said to employ **classical statistics.** However, if the molecules were indistinguishable, then finding zero, one, two, or three molecules on the left side would be equally probable (probability = ¼). This is an example of **quantum statistics.**

Two kinds of quantum statistics are required because when particle wave functions overlap, interchanging the particles can result in either leaving the wave function for the system unchanged or changed by an overall minus sign. If the wave function is unchanged, then it's said to be symmetric, and the particles follow the **Bose–Einstein distribution.** If the wave function is changed by a minus sign, then it's said to be antisymmetric, and the particles follow the **Fermi–Dirac distribution.** In the "classical limit" of nonoverlapping wave functions, both kinds of quantum statistics reduce to the classical distribution for distinguishable particles.

There's a correlation between a particle's spin and its quantum distribution. Particles including electrons, protons, and neutrons, which have half-integer spin, follow the Fermi–Dirac distribution and are called **fermions.** Particles that have zero or whole-integer spin, such as photons and many whole atoms, follow the Bose–Einstein distribution and are called **bosons.**

Fermions and bosons have radically different behavior. Fermions obey the Pauli exclusion principle (Chapter 4), which means that no two fermions may have the same wave function or same set of quantum numbers. This restriction forces electrons in an atom to occupy subshells, with each subshell having limited occupancy. Outer (valence) electrons in unfilled subshells are responsible for the atom's chemical behavior. Further, the observed electrical conductivity of conductors and semiconductors is a result of their adherence to the Pauli principle.

Bosons have no such restriction; therefore, at extremely low temperatures, some materials made exclusively of bosons exhibit **Bose–Einstein condensation,** in which a large number of particles collapse into a single quantum state. For example, at temperatures below 2.2 K, liquid helium enters a superfluid phase in which it exhibits properties of a Bose–Einstein condensate and flows without viscosity. Physicists first observed Bose–Einstein condensation in gases cooled to 1 μK and lower in the 1990s.

HOW COLD CAN YOU GET?

Physicists have long sought to achieve lower and lower temperatures, not only for practical reasons but also to see how materials behave when they are much colder than we normally experience. In the early twentieth century the Dutch physicist Heike Kamerlingh Onnes liquefied helium at 4.2 K (at 1 atm pressure), the lowest boiling point known for any material. This enabled Onnes to discover superconductors—materials that conduct with absolutely no electrical resistance at a low enough temperature (Chapter 2). Lowering the pressure of liquefied helium lowers its boiling point to as low as 1 K, and in that process it's observed that helium becomes a superfluid below 2.2 K (Chapter 3). A superfluid is an example of Bose–Einstein condensation.

Achieving temperatures much below 1 K requires more sophisticated techniques. A common method known as **adiabatic demagnetization** uses a paramagnetic salt that is first cooled to as low a temperature as possible in liquid helium. Then an applied magnetic field aligns the paramagnet's dipoles while the low temperature is maintained. Reducing the applied magnetic field allows the magnetic dipoles to randomize, increasing their entropy. If this is done adiabatically (i.e., the material is insulated from its environment), there is a corresponding decrease in thermal entropy, and the sample cools. Adiabatic demagnetization allows physicists to achieve temperatures on the order of millikelvins, or even microkelvins if nuclear magnetic moments are used.

Laser cooling is another method that can achieve ultracold temperatures. In this method a laser is fired into a gas of atoms already cooled to a low temperature. The laser's photon energy is tuned just below the energy that will excite an atomic transition in the gas's atoms. However, for atoms moving toward the laser, the incoming photons are Doppler-shifted to a higher energy that will excite the transition. Conservation of momentum requires the atom absorbing the incoming photon to slow down, resulting in a net loss of energy for the gas.

The effect after many photon absorptions is a lower gas temperature, routinely in the microkelvin range or lower.

In 2003 a research team at MIT achieved a temperature of 450 pK (= 4.5 × 10^{-10} K) using sodium atoms. This experiment involved laser cooling followed by a series of other methods designed to eliminate faster atoms from the sample, leaving behind only the slowest ones.

FURTHER READINGS

Assael, Marc J., Goodwin, Anthony R. H., Stamatoudis, Michael, Wakeham, William A., and Will, Stefan 2011. *Commonly Asked Questions in Thermodynamics*. Boca Raton FL: CRC Press, Taylor & Francis.

Atkins, Peter 2010. *The Laws of Thermodynamics: A Very Short Introduction*. Oxford, England: Oxford University Press.

Baierlein, Ralph 1999. *Thermal Physics*. Cambridge, England: Cambridge University Press.

Schroeder, Daniel V. 2000. *An Introduction to Thermal Physics*. San Francisco, CA: Addison Wesley Longman.

Chapter 7

Atoms and Nuclei

One of the oldest problems in physics—and in all of science—is the question of what everything is made of. There's a natural curiosity about this that children express and that is seen throughout world cultures. The approach we have followed for many centuries is to try to understand matter by breaking it into smaller and smaller pieces, looking for fundamental building blocks. This search has led to our understanding of the atomic nature of matter. That is, all matter we normally see or encounter is made up of atoms, with the wide variety of matter explained by different combinations of atoms. More in-depth study of atoms has shown that they are not solid bodies, but instead are made up of a small atomic nucleus plus electrons. In the last century, research on the nucleus has led to a deeper understanding of its nature and a variety of applications—some useful and others terrifying. In this chapter we'll address some of the fundamental ideas about atoms and nuclei.

 ATOMS AND ELEMENTS—WHAT ARE THEY?

We're all naturally curious about the world around us. Where do things come from, and what makes them work the way they do? The visible is often governed by the invisible. A child with a toy may want to look inside or take the toy apart to see how it works. The idea of **atoms** is something like that: Break matter into smaller and smaller pieces, and eventually you'll find some fundamental pieces out of which everything is made.

That idea goes back at least as far as the ancient Greeks, who believed that there were just a few types of atoms, with each individual atom being indivisible and unchangeable. (The word "atom" comes from Greek roots meaning, literally, "that which cannot be cut.") The Greeks conceived all matter to consist of different combinations of these few atomic types. In one popular model, there were four types: air, earth, fire, and water. Our modern conception is similar to the ancient one but different in some important ways. For one thing, what we now call atoms are not single indivisible entities, but rather composite, made up of protons, neutrons, and electrons. Further, the protons and neutrons are joined together in a small atomic nucleus, with electrons found outside the nucleus.

Atoms are categorized by **element,** with each atom of a given element having the same number of protons in its nucleus. For example, all atoms of the element hydrogen have a single proton in their nucleus, with a charge of $+e = +1.60 \times 10^{-19}$ C. Similarly, helium has two protons in its nucleus and nuclear charge $+2e$, lithium three protons and charge $+3e$, and so on. There are 90 naturally occurring elements, though we normally encounter only several dozen of them in everyday life. Carbon, hydrogen, and oxygen are the basic elements found in the organic molecules that make up living things. Many other elements are essential to life as we know it, including nitrogen, sodium, iron, and potassium. Read the label of your multivitamin, and you'll see more of them! Other elements, such as copper and aluminum, are found extensively in industrial applications and consumer goods.

Within the last few centuries, scientists expanded their knowledge about the elements and chemistry, identifying more elements and discovering their properties. In 1869 Russian chemist Dmitri Mendeleev created the first **periodic table** of the elements, with elements ordered by atomic weight and arranged in groups based on their chemical properties. At the time, however, it was not understood why elements in a certain group—for example, fluorine, chlorine, bromine, and iodine—had similar chemical properties. This would require a better understanding of atomic structure, which came in the early twentieth century.

What Are Atomic Spectra?

You've probably seen differently colored flames coming from materials as they burn, as well as fluorescent lights with various colors—from the pale blue of fluorescent lights to the bright red of neon. **Atomic spectroscopy** is the study of the connection between these colors and the atoms that produce them, and spectroscopy grew in the nineteenth century as a new way to study atoms. It was found that elements exhibited characteristically colored spectra when heated or when reacted with other elements. Notably, helium was discovered in the solar spectrum before it was found on Earth. The noble gas helium does not react easily with other elements and, because of its low density, helium easily escapes Earth's atmosphere.

WHAT IS THE RUTHERFORD–BOHR MODEL OF THE ATOM?

Knowing the properties of atoms is good, but there's much more to be known. What's inside the atom, and what exactly is changing in atoms when they interact with each other? An important step was taken in 1897, when J. J. Thomson

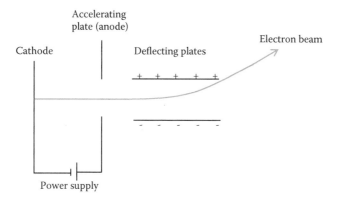

Figure 7.1   J. J. Thomson's charge-to-mass experiment. Electrons are accelerated from the cathode at the far left and travel between the charged plates. The electrons are deflected toward the positive plate, with the amount of deflection revealing e/m.

measured the **charge-to-mass ratio** for electrons. He did so using the device shown schematically in Figure 7.1, consisting of an evacuated glass tube with two metal electrodes. In this experiment, a potential difference is established between the electrodes, with the lower potential cathode and the higher potential anode. Heating the cathode releases electrons—or "cathode rays," as they were called at the time. The electrons pass through a pair of parallel plates, which also have some potential difference established between them. The resulting electric field between the parallel plates causes the electron beam to deflect, and measuring the amount of deflection reveals the electron's charge-to-mass ratio e/m.

Thomson's 1897 experiment yielded results that were slightly lower than the current accepted value $e/m = 1.76 \times 10^{11}$ C/kg, but good enough to reveal that the electron's charge-to-mass ratio was much higher than that of ionized hydrogen, the smallest atom, which has $e/m = 9.58 \times 10^7$ C/kg. This indicates that the negatively charged electrons must be much less massive than the positive part of the atom—a fact for which any model of the atom needs to account.

Another important piece of evidence came from the lab of Ernest Rutherford in England. There, Hans Geiger and Ernest Marsden carried out a series of experiments in 1909 in which they bombarded extremely thin gold foil with **alpha particles** (doubly ionized helium). Most of the energetic alpha particles passed through the foil with little deflection, but some experienced larger deflections, with a few even bouncing straight backward (Figure 7.2). Rutherford was at first surprised by the larger deflections because of the high kinetic energy of the alpha particles and the thinness of the foil. After analyzing the results, Rutherford concluded in 1911 that the only explanation was that the massive

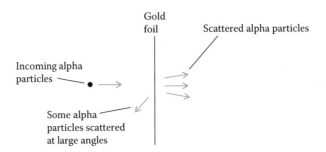

Figure 7.2   Schematic representation of the alpha scattering experiment used to demonstrate the existence of the atomic nucleus. Most alpha particles pass through the gold foil with little deflection, but some are deflected significantly. Analysis of the results is consistent with the existence of a dense, positively charged nucleus.

positive part of the atom must be contained in a very small, dense nucleus with a radius on the order of 10^{-15} m, compared with the atom's radius on the order of 10^{-10} m. We'll say a lot more about the nucleus later in this chapter.

The next step in atomic theory was taken by Danish physicist Niels Bohr, who worked with Rutherford at Cambridge beginning in 1912. Bohr created a model in which negative electrons orbit around the massive, positive nucleus, analogous to the way in which planets orbit the sun. Physicists were reluctant to accept such a model because orbits require accelerated motion and, according to classical electromagnetic theory, accelerated charges must radiate energy continuously. Bohr explained that the orbiting electrons would not radiate energy if the orbits were of certain fixed sizes, governed by Planck's quantum constant h. Thus, the stable orbits are said to be **quantized.**

GOING DEEPER—INSIDE THE BOHR ATOM

The stable radii found by Bohr are

$$r = n^2 a_0 \tag{7.1}$$

where n is a positive integer (1, 2, 3, etc.) and the Bohr radius a_0 is given by

$$a_0 = \frac{4\pi\varepsilon_0\hbar^2}{me^2} \tag{7.2}$$

In Equation (7.2), m and e are the electron's mass and the magnitude of its charge, and $\hbar$ is Planck's constant h divided by 2π. The Bohr radius is the smallest radius of a hydrogen atom, corresponding to the $n = 1$ orbit, and has numerical value $a_0 = 5.29 \times 10^{-11}$ m, in good agreement with the known size of hydrogen.

Bohr's semiclassical analysis of the orbit gave, for the energy of a hydrogen atom,

$$E = -\frac{e^2}{8\pi\varepsilon_0 a_0 n^2} = -\frac{13.6 \text{ eV}}{n^2} \tag{7.3}$$

Together, Equations (7.1) and (7.3) provide a good description of the Rutherford–Bohr atom. Some of the lower level electron orbits are shown in Figure 7.3. The real test of the model comes from measuring the energy levels (Equation 7.3), which is done by examining the emission spectrum of hydrogen (Figure 7.4). If the model is correct, when a hydrogen atom makes a transition from a higher to lower energy level, the corresponding energy is released in the form of a quantum of light, called a **photon.** According to Planck's quantum theory, the relationship between the photon's energy E and wavelength λ is

$$E = \frac{hc}{\lambda} \tag{7.4}$$

A transition from the $n = 3$ level to $n = 2$ is predicted by Equations (7.3) and (7.4) to yield a photon of wavelength 656 nm, which matches the wavelength in the observed spectrum (Figure 7.4). Other transitions between energy levels yield observed spectral lines in agreement with Bohr's theory, including those lines in the infrared and ultraviolet parts of the spectrum. For example, the transition from $n = 2$ to $n = 1$ produces a 121-nm ultraviolet photon.

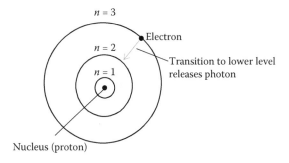

Figure 7.3 Bohr's model of hydrogen (not to scale). When an electron makes a transition from a higher to lower level, a photon is emitted, with wavelength given by Equation (7.4).

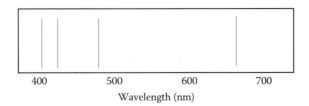

Wavelength (nm)

Figure 7.4   The four visible spectral lines for hydrogen. The wavelengths match those predicted by Bohr's quantum theory.

WHAT ARE ATOMIC ORBITALS AND SHELLS?

Unfortunately, the Rutherford–Bohr model can't be applied to atoms larger than hydrogen that have more than one electron because the significant electron–electron interaction can't be taken into account. However, the energy levels $n = 1, 2$, etc. in this model, along with the quantum-mechanical description of the hydrogen atom (Chapter 4), provide a framework for explaining the behavior of other atoms.

An electron in a hydrogen atom is described by four **quantum numbers:**

$n = 1, 2, 3$, etc.—principal quantum number

$l = 0, 1, 2, 3$, etc. (with restriction $l < n$)—orbital angular momentum quantum number

$m_l = -l, -l + 1, ..., 0, 1, ... l - 1, l$ (with restriction $|m_l| \leq l$)—magnetic quantum number

$m_s = \pm \frac{1}{2}$—spin quantum number

The four quantum numbers result from the full quantum-mechanical treatment of the atom. The first quantum number (n) corresponds to the Rutherford–Bohr quantum number n in hydrogen. In hydrogen, the single electron has a set of four quantum numbers (n, l, m_l, m_s), subject to the restrictions listed previously, which together describe the quantum state of the atom. Examples of valid sets of quantum numbers are $(1, 0, 0, \frac{1}{2})$ and $(3, 2, -1, -\frac{1}{2})$.

The basic properties of atoms follow from the combination of quantum numbers n and l, which defines the atomic **orbital.** In the spectroscopic notation used to describe electronic states, the letters s, p, d, and f are used (respectively) to describe the states with $l = 0, 1, 2, 3$. Thus, the combination $n = 1, l = 0$ represents a 1s orbital, and $n = 4, l = 2$ represents a 4d orbital.

For multielectron atoms, electrons fill in order of increasing energy, according to several empirical rules. The Pauli exclusion principle states that no two electrons may have the same set of four quantum numbers.

This restricts the number of electrons in a particular orbital. For example, in an s-orbital ($l = 0$), the restriction $|m_l| \leq l$ requires that $m_l = 0$. With two possible values of m_s, there can be no more than two electrons in a particular s-orbital. Similarly, the limits on p-, f-, and d-orbitals are 6, 10, and 14 electrons, respectively.

In multielectron atoms, a particular n level is called a **shell,** and the nl combination (e.g., 3p) is a **subshell.** With certain exceptions, electron subshells fill one at a time in an order given by Madelung's rules, which state that (1) subshells fill in order of increasing $n + l$, and (2) when values of $n + l$ are equal, the lower value of n is preferred. With these rules in mind, the resulting order of filling is 1s, 2s, 2p, 3s, 3p, 4s, 3d, 4p, 5s, 4d, 5p, 6s, 4f, 5d, 6p, 7s, 5f, 6d, 7p.

The full electron configuration of an atom is described by the number of electrons in each subshell. Hydrogen has one 1s electron and is described as $1s^1$. Helium has two 1s electrons and is therefore $1s^2$. This fills the 1s subshell, so the third element, lithium, is $1s^2 2s^1$. Knowing the number of electrons and the order of subshell filling normally allows you to write the electron configuration. For example, aluminum, with 13 electrons, is $1s^2 2s^2 2p^6 3s^2 3p^1$.

There are a few exceptions to the principle of filling one subshell at a time. For example, nickel, with 28 electrons, has outer subshells $4s^2 3d^8$. (For larger atoms, it's customary to list just the outer subshells, to avoid the tedious exercise of listing lower subshells that are obviously full.) You might expect copper, with 29 electrons, to be $4s^2 3d^9$, but this is incorrect. Filling the d-subshell results in a more stable configuration, so copper becomes $4s^1 3d^{10}$. This configuration leaves the single 4s electron weakly bound, which makes copper an excellent conductor of electricity.

How Do Atomic Shells and Subshells Explain the Periodic Table?

Understanding the subshell structure of atoms explains the groups of chemically similar elements found in the periodic table. The **halogens** all have five electrons in a p subshell: fluorine = $2p^5$, chlorine = $3p^5$, bromine = $4p^5$, and iodine = $5p^5$. As a result, the halogens tend to form negative ions by accepting an electron to fill the p-orbital, and they readily form compounds with metals such as sodium and potassium that have a single s-electron, as in sodium chloride NaCl. **Noble gases** (helium, neon, and so on) have filled p-subshells, making them highly resistive to reacting with any other elements. **Transition elements** (found in the middle of the periodic table) have unfilled d-subshells. The tendency of unpaired d-electrons to align their spins makes them good candidates for ferromagnetic and paramagnetic behavior.

HOW ARE X-RAYS PRODUCED?

You've probably had medical x-ray photographs taken of your body—most likely dental x-rays and maybe pictures of a broken bone if you've been unlucky! The atomic transitions we've been discussing explain how x-rays are made.

The optical spectra exhibited by many atoms result from transitions like the ones in hydrogen but involving electrons in the outermost shells. The energy difference between adjacent outer shells of a multielectron atom are generally on the order of electron volts (eV), which results in the release of a photon having wavelength on the order of 10^{-7} m. Such photons are either in the visible range, the near ultraviolet, or near infrared. Figure 7.5 shows the visible spectrum for helium. Compared with the hydrogen spectrum shown in Figure 7.4, this pattern of spectral lines appears fairly random. Hydrogen is the only element with energy levels that follow a regular mathematical pattern—the one given in Equation (7.3)—so the optical spectra for other elements do not follow a regular pattern.

X-rays are also produced by electron transitions in multielectron atoms, but they come from transitions to the inner shells, usually the first and second. The available energy in such transitions increases with the number of protons in the nucleus. Thus, while a transition from the $n = 2$ shell to the $n = 1$ shell in hydrogen produces a photon of about 10 eV, the corresponding transition in a larger atom can be on the order of kiloelectron volts, yielding an x-ray photon with a wavelength of 1 nm or shorter.

Remember that the lower subshells are normally full in multielectron atoms. For this reason, some effort is required to dislodge an inner-shell electron and allow an x-ray transition to occur. The device shown in Figure 7.6 is commonly used to generate x-rays. Electrons from a cathode are accelerated to high energies and directed into a solid (usually metal) target. The energy of the incoming electron must exceed the binding energy of the inner-shell electron in order to remove that electron from the atom. Once this happens, an electron falling from a higher shell to fill the lower shell vacancy generates the x-ray.

X-rays are categorized by the transition that creates them. An old nomenclature referred to the $n = 1, 2, 3...$ shells by letters K, L, M,...; using this notation,

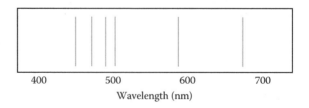

Figure 7.5 The visible spectrum for helium.

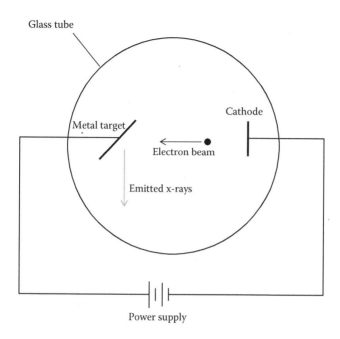

Figure 7.6 Schematic of an x-ray tube.

x-rays are named by the shell where the transitioning electron lands. A transition from $n = 2$ to $n = 1$ produces a K_α x-ray. The $n = 3$ to $n = 1$ transition produces a K_β x-ray, and so on. Similarly, the $n = 3$ to $n = 2$ transition produces an L_α x-ray.

In 1913, Harry Moseley made a systematic study of x-rays produced by different elements. He found that x-ray energy increased systematically with atomic number Z. In addition to explaining the source of x-rays and confirming the shell model of the atom, Moseley's work led to a rearrangement of the periodic table in some places. Until that time, chemists ordered the elements by atomic mass. Generally, atomic mass increases with atomic number, but there are exceptions. One example is nickel $(Z = 28)$, which has a slightly lower atomic mass than cobalt $(Z = 27)$. Ordering the periodic table by atomic number is consistent with the ordering by atomic subshell, as given earlier.

What Is the Difference between Characteristic and Bremsstrahlung X-Rays?

To this point we have been describing **characteristic x-rays** produced by atomic transitions. Accelerated electrons can also give up their energy gradually when they interact with matter, resulting in a continuous x-ray spectrum.

This process is known as **Bremsstrahlung.** Often x-ray spectra contain both characteristic and Bremsstrahlung x-rays, appearing as a continuous spectrum with spikes of intensity at the characteristic wavelengths.

How Do Medical X-Rays Work?

X-rays that enter your body are scattered or absorbed at different rates by different kinds of tissue. An x-ray source sends the x-rays through part of your body, and they are received by a detector on the other side. There's a distinct difference between absorption by bone and soft tissue, so x-rays are particularly useful in imaging bones (Figure 7.7). However, they can also reveal abnormalities in your lungs and other organs.

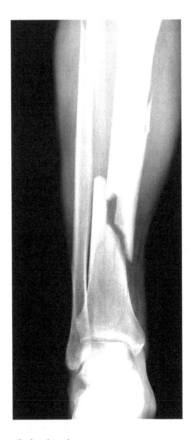

Figure 7.7   X-ray image of a broken bone.

WHAT ARE THE BASIC PROPERTIES OF NUCLEI?

The alpha-particle scattering experiments done by Rutherford's group in the early twentieth century revealed the nucleus as a small, dense mass containing the atom's positive charge and almost all of its mass, with the remaining mass in the electrons that move outside the nucleus. However, such experiments couldn't detect the internal structure of the nucleus. Physicists puzzled about this until 1932, when James Chadwick discovered the **neutron,** a neutral particle just slightly more massive than the proton. (In atomic mass units, the proton's mass is 1.00728 u, and the neutron's mass is 1.00866 u.) It had been known for several years that alpha particles bombarding beryllium release an energetic beam of neutral particles or radiation. By studying the effects of this neutral beam on other materials, Chadwick deduced that it consisted of neutral particles with a mass of about 1 u, which he recognized as the neutron.

Physicists soon realized that the nucleus is made up of protons and neutrons. This is consistent with the observation that atoms all have atomic masses that are fairly close to an integer number of atomic mass units. Each nucleus consists of Z protons and N neutrons, where Z is the atomic number. The electric charge of the nucleus of Z protons is $+Ze$. Protons and neutrons in a nucleus are referred to collectively as **nucleons,** and the total number of nucleons $A = Z + N$ is called the **mass number.**

Each element is characterized by Z, the number of protons it contains. Atoms with $Z = 6$ are all called carbon, regardless of the number of neutrons they contain. Atoms with the same Z but different N are called **isotopes,** and each specific combination of Z and N is called a **nuclide.** A compact notation used to identify each nuclide is $^{A}_{Z}X$, where X is the chemical symbol of the element with Z protons and mass number A. Thus, the common isotope of carbon that contains six protons and six neutrons is $^{12}_{6}C$, and the isotope of carbon with eight neutrons is $^{14}_{6}C$. The Z subscript is redundant and can be omitted (as in ^{12}C), but the subscript may be included when writing reactions so that the number of protons in each nuclide is evident.

Nuclei can be thought of as a close packing of protons and neutrons in a roughly spherical shape (Figure 7.8). (Think of a bag packed full of soccer balls, with each ball representing a proton or neutron.) Because protons and neutrons are about the same size, the volume of the nucleus grows in proportion to A, the number of nucleons.

WHAT IS THE STRONG (NUCLEAR) FORCE?

The positively charged protons packed closely together experience mutual repulsion due to their electric charge. The inverse-square nature of the Coulomb force makes the repulsion quite intense at such short distances. For example,

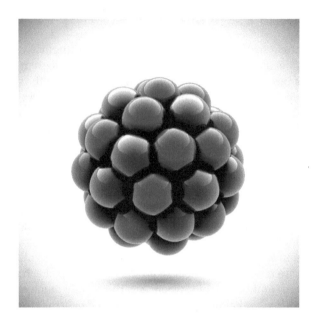

Figure 7.8   Model of a nucleus as a closely packed sphere of protons and neutrons.

two protons separated by a distance of 3 fm experience a repulsive force of about 25 N—a huge force for such small particles! This implies that there must be a force even stronger than electromagnetism responsible for binding the nucleus together. That attractive force is called the strong force.

The strong force acts between **hadrons,** which are composite particles made of quarks. Hadrons come in two types: **baryons,** which are composed of three quarks, and **mesons,** which consist of a quark–antiquark pair. The strong force is responsible for binding the quarks together in these composite particles. We'll consider the properties of quarks, baryons, and mesons in Chapter 8.

At the femtometer scale on which the strong force is effective, it is typically about two orders of a magnitude stronger than the electromagnetic force. (That's why it's called the strong force!) That extreme strength allows quarks to be bound into hadrons and baryons, and it allows baryons to be bound to each other, even if they are repelled from one another by their charge. Specifically, protons and neutrons (which are both baryons) are bound together by the strong force to form nuclei. When the strong force acts between protons and neutrons in a nucleus, it is commonly referred to as the **nuclear force.**

HOW ARE NUCLEI BOUND BY THE NUCLEAR FORCE?

The nuclear force acts between any pair of nucleons, whether proton–proton, proton–neutron, or neutron–neutron. Protons repel each other electromagnetically, but the net effect of the nuclear forces between all the nucleons overcomes the electromagnetic force to form stable nuclei. The nuclear force is extremely short range, falling off sharply with distance and effectively reaching zero at distances of 3 fm and larger. Although the electromagnetic force drops off with distance according to the inverse-square Coulomb law, it is still present at distances of 3 fm and larger. That's why nuclear stability becomes harder to achieve as nuclear size increases. There are no stable nuclei larger than ^{209}Bi.

You can think of the binding of a nucleus as a competition between the attractive nuclear force and the repulsive electromagnetic force. The other two forces in nature are the weak and gravitational forces. The **weak force,** which will be discussed in more detail in Chapter 8, governs beta decay in the nucleus, and therefore it does play a role in nuclear stability. However, the weak force doesn't contribute to the attraction between nucleons. The gravitational force is present between all particles with mass, but because it's some 40 orders of magnitude weaker than the nuclear force, gravity also plays no role in nuclear binding.

Unstable nuclides vastly outnumber stable ones. The key to nuclear binding and stability is the ratio N/Z of neutrons to protons. For smaller nuclei, the ratio must be close to 1 to obtain a stable nuclide. (It's exactly 1 for many common isotopes, such as ^{4}He, ^{12}C, and ^{16}O.) The ratio N/Z required for stability increases gradually for heavier nuclides, exceeding 1.5 for ^{209}Bi. (There are no stable nuclides beyond $Z = 83$.) This behavior is explained by considering the growing strength of electromagnetic repulsion as protons are added to a stable nucleus. Adding relatively more neutrons to the mix allows for more strong-force binding to take place among the protons and neutrons, while at the same time spacing out the protons to reduce the net electromagnetic repulsion.

Which Combinations of Z and N Are More Stable?

Although the nuclear force is roughly the same strength between any pair of nucleons, **even–even** isotopes (with even numbers of both protons and neutrons) tend to be the most stable. Protons and neutrons are both fermions, which obey the Pauli exclusion principle (Chapter 4). Thus, a pair of protons can have opposite spins but otherwise be in the same quantum state. The same goes for a pair of neutrons. Such a situation is generally more stable than any other combination of protons and neutrons with the same mass number. **Odd–odd** nuclides (with odd numbers of both protons and neutrons) are particularly unstable. In fact, there are only four of them that are stable, and none is heavier than ^{14}N. The other combinations (**even–odd** and **odd–even**) tend to be intermediate in stability.

WHAT IS NUCLEAR BINDING ENERGY?

Nuclear binding energy is the net attractive energy of an entire nucleus. Conversely, the binding energy is the amount of energy you would need to supply to break the nucleus into free neutrons and protons. Conservation of mass–energy gives a simple expression for the binding energy of a generic nuclide $^{A}_{Z}X$ having atomic number Z and mass number A. The observed binding energy results from the mass difference between the nuclide and its constituent Z protons plus $N = A - Z$ neutrons. The binding energy is that mass difference multiplied by c^2.

Binding energy increases as the mass number of a nuclide increases, so the best way to compare binding energies of different nuclides is to compute the **binding energy per nucleon** B/A, as illustrated in Figure 7.9. Nuclei in the middle of the periodic table tend to be most stable, with the peak at ^{56}Fe. This means that ^{56}Fe is the most stable isotope, relatively speaking. Notice that it's an even–even nuclide, as are several other nuclides that appear as distinctive peaks in Figure 7.9, such as ^{4}He and ^{16}O.

WHAT ARE THE BENEFITS AND DANGERS OF RADIOACTIVE NUCLEI?

If you've ever been around radioactive materials, you've probably seen them accompanied by big signs warning you to keep away from the danger. Radiation doses can be extremely harmful, even fatal. But when controlled and used the right way, radiation can extend life by being used to diagnose and treat diseases.

Radioactive decay is a spontaneous change in a nuclide by the emission of a particle or photon. The three common forms of radioactive decay are **alpha decay,** in which the nucleus emits an alpha particle (helium nucleus); **beta decay,** which involves emission or capture of an electron or emission of a positron (the positive antiparticle of the electron); and **gamma decay,** the emission of a high-energy photon. In each case, the decay occurs because the resulting system of particles is more stable than the original nuclide.

What Is Alpha Decay?

Alpha decay is the emission of an alpha particle (helium nucleus). Many radioactive nuclides decay via this mode, due to the relatively high stability of the

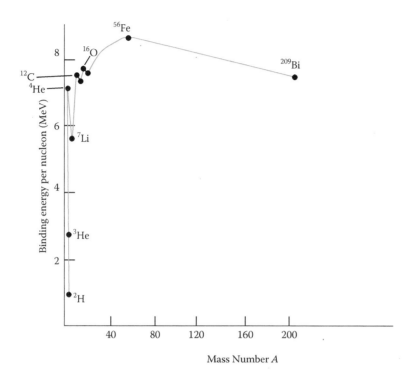

Figure 7.9   Graph of binding energy per nucleon. Notice the peak at ^{56}Fe and spikes for particularly stable nuclei, such as ^{4}He.

^{4}He nuclide (see Figure 7.9). The binding energy of the ^{4}He nucleus is 28.3 MeV, so any nuclide that has its last two protons and two neutrons bound by less than 28.3 MeV is subject to alpha decay.

Because the helium nucleus carries away two protons and two neutrons, a generic expression for the alpha decay of a nuclide X is

$$\begin{array}{l}^{A}_{Z}X \rightarrow\,^{A-4}_{Z-2}D + \,^{4}_{2}He\end{array}$$ (7.5)

The symbol D is used for the generic nucleus that results from a radioactive decay because it is often called the **daughter** nucleus, with X known as the **parent**. The energy Q released in the alpha decay is due to the difference in masses before and after the decay, per Einstein's mass–energy relation:

$$Q = \left[M\left(^{A}_{Z}X\right) - M\left(^{A-4}_{Z-2}D\right) - M\left(^{4}_{2}He\right) \right]c^{2}$$ (7.6)

Alpha decay can occur for a nuclide X if $Q > 0$ for that nuclide in Equation (7.6). $Q > 0$ means that the resulting daughter and alpha particles are more stable than the original nuclide. When alpha decay occurs, Q is the excess energy that is released in the form of kinetic energy of the products (daughter and alpha particles). Generally, the alpha is much lighter than the daughter, and therefore conservation of momentum requires that the alpha carry the bulk of the kinetic energy. For example, the alpha decay of ^{230}U has $Q = 5.99$ MeV and $K_\alpha = 5.89$ MeV.

Most alpha emitters are heavier nuclides with a relatively low ratio of neutrons to protons (N/Z). Emission of an alpha particle (two protons and two neutrons) results in a nuclide with a slightly higher N/Z, which is more likely to be stable. This works for more massive nuclei, but for lighter nuclei there is little effect because it doesn't help approach the favorable ratio $N/Z = 1$. Therefore, lighter nuclides tend not to alpha decay.

What Is Beta Decay?

Beta decay comes in three forms and refers to the beta particle, which is an electron or positron. In the context of beta decay, the electron (a negative particle) is designated β^-, and the positron is β^+. The positron has the same mass as the electron and is its antiparticle. The three forms of beta decay are emission of a β^-, emission of a β^+, and capture of an electron. Just as for alpha decay, it's possible to write a generic reaction and compute a Q value for each form of beta decay. For β^- emission:

$$_{Z}^{A}X \rightarrow _{Z+1}^{A}D + _{-1}^{0}\beta^- + \overline{\nu} \qquad Q = \left[M\left(_{Z}^{A}X\right) - M\left(_{Z+1}^{A}D\right)\right]c^2 \quad (\beta^- \text{ emission}) \quad (7.7)$$

Notice that in addition to the β^-, another particle is emitted: an **antineutrino** (symbol $\overline{\nu}$), which carries away some of the energy of the decay. Neutrinos and antineutrinos will be discussed in more detail in Chapter 8, along with the weak force, which is responsible for the stability of nuclides in beta decay.

The reason the emitted electron's mass doesn't appear in the computation of Q is that a table of *atomic* masses includes the Z electrons in the atom. Because the daughter D has $Z + 1$ electrons, the emitted electron's mass is already included. The antineutrino's mass is less than 4×10^{-36} kg (compared with the electron's mass of 9×10^{-31} kg) and is too small to be included in the expression for Q. The relatively heavy nucleus suffers very little recoil (compared with alpha decay), so almost all of the energy Q shows up in the kinetic energies of the emitted electron and antineutrino.

A common β^- emitter is ^{14}C, which occurs naturally. The reaction and Q value from Equation (7.7) are

$$_{6}^{14}C \rightarrow _{7}^{14}N + _{-1}^{0}\beta^- + \overline{\nu} \qquad Q = 0.16 \text{ MeV}$$

A free neutron is also a beta emitter, via the reaction ${}_{0}^{1}n \rightarrow {}_{1}^{1}H + {}_{-1}^{0}\beta^{-} + \bar{\nu}$. The resulting hydrogen nucleus is just a proton. In a sense, the β^- decay always involves the conversion of a neutron to a proton within the nucleus, consistent with the reaction in Equation (7.7). It's a curious result of quantum mechanics that neutrons within a nucleus can remain stable indefinitely, but free neutrons outside the nucleus decay in this reaction within a short time.

Typically, β^- emitters are the relatively neutron-rich isotopes of their elements. A β^- decay decreases the number of neutrons and increases the number of protons (by one each), moving the daughter product closer to a stable isotope.

By reasoning similar to that used to find Equation (7.7), the corresponding relations for positron decay are

$$ {}_{Z}^{A}X \rightarrow {}_{Z-1}^{A}D + {}_{1}^{0}\beta^{+} + \nu \quad Q = \left[M\left({}_{Z}^{A}X\right) - M\left({}_{Z-1}^{A}D\right) - 2m \right]c^{2} \quad (\beta^{+} \text{ emission}) \quad (7.8) $$

Now the additional particle emitted is a **neutrino** (ν). The computation of Q requires two electron masses m, one for the emitted β^+ and the other to account for the fact that the daughter atom listed in the mass table has one fewer electron than the parent. In general, β^+ emitters are relatively neutron-poor isotopes of their elements. The β^+ decay is the equivalent of the conversion of a proton to a neutron within the nucleus, which tends to make the daughter more stable. However, unlike the corresponding process in β^- decay (conversion of a neutron to a proton), free protons do not decay.

For electron capture, the reaction and Q-value are

$$ {}_{Z}^{A}X + {}_{-1}^{0}\beta^{-} \rightarrow {}_{Z-1}^{A}D + \nu \quad Q = \left[M\left({}_{Z}^{A}X\right) - M\left({}_{Z-1}^{A}D\right) \right]c^{2} \quad (\text{electron capture}) \quad (7.9) $$

Electron capture and positron emission are similar in that, for a given radioactive nuclide, they result in the same daughter nucleus. One difference is that when an inner-shell electron is captured, the resulting vacancy in that shell leads to the emission of a characteristic x-ray. Also, the Q-values for the two processes differ by two electron masses, or about 1 MeV, so there will be some radioactive nuclides for which electron capture is possible but not positron emission. For example, ^{55}Fe has $Q = +0.23$ MeV for electron capture, so that process is allowed, but β^+ emission is not allowed because $Q = -0.79$ MeV.

What Is Gamma Decay?

Gamma decay is the emission of a gamma-ray photon by a nucleus. The photon is uncharged and does not involve the removal of nucleons or the changing of nucleons into other particles, so the gamma-ray emission is described by

$$\,_Z^A X \rightarrow \,_Z^A X + \gamma \qquad\qquad (7.10)$$

where γ is the gamma ray.

The source of the energy is a transition of the nucleus from a higher energy state to a lower one. The released energy Q is due to the energy difference between nuclear states, with most of the decay energy going to the photon and the rest to the recoiling nucleus, as required by conservation of momentum. A gamma decay is often the secondary result of an alpha or beta decay, which may leave the daughter nucleus in an excited state. Other processes, including neutron bombardment, fission, and fusion, may also create an excited nucleus that will subsequently undergo gamma decay.

What Is a Radioactive Half-Life?

Each radioactive nuclide decays at a different rate. Some decay very quickly, and others more slowly. A nuclide's **half-life** is the time it takes for half of a sample of that nuclide to decay.

The radioactive decay process is a random one. Any given radioactive nucleus has a certain probability of decaying within a given time period, but it's impossible to say when that nucleus will decay. For that reason, half-life measurements for a particular nuclide vary from one trial to another. Accepted half-life values have been determined through the statistical analysis of many trials. Because of the statistical nature of the decay, larger sample sizes give better results.

Half-lives vary from small fractions of a second to billions of years. For example, the half-life of ^{213}At is 125 ns, while the half-life of ^{238}U is 4.47×10^9 y. The half-life of ^{238}U is comparable to the age of our solar system. That explains why it is found naturally, even though it is radioactive. Although ^{209}Bi is the heaviest stable nuclide, there are many other long-lived heavier nuclides that are found naturally. Three of them are uranium: ^{234}U, ^{235}U, and ^{238}U. The relative abundance of the three isotopes is due to their different half-lives, with longer half-lives corresponding to greater abundance. Those abundances are 0.0055% for ^{234}U ($t_{1/2} = 2.46 \times 10^5$ y), 0.72% for ^{235}U ($t_{1/2} = 7.04 \times 10^8$ y), and 99.27% for ^{238}U ($t_{1/2} = 4.47 \times 10^9$ y).

What Are Decay Chains?

When a radioactive nucleus decays, there is no guarantee that the daughter nucleus will be stable. Although each decay leads to a more stable configuration, it may take multiple decays before complete stability is achieved. A sequence of decays leading to a stable nuclide is called a decay chain.

Decay chains are particularly relevant for heavier nuclides. Because ^{209}Bi is the heaviest stable nuclide, any heavier nuclide will eventually decay. This must

TABLE 7.1   RADIOACTIVE SERIES

Mass Numbers	Series Name	Common Parent	Total $t_{1/2}$ (y)	Stable Product
$4n$	Thorium	$^{232}_{90}\text{Th}$	1.40×10^{10}	$^{208}_{82}\text{Pb}$
$4n + 1$	Neptunium	$^{237}_{93}\text{Np}$	2.14×10^{6}	$^{209}_{83}\text{Bi}$
$4n + 2$	Radium (uranium)	$^{238}_{92}\text{U}$	4.47×10^{9}	$^{206}_{82}\text{Pb}$
$4n + 3$	Actinium	$^{235}_{92}\text{U}$	7.04×10^{8}	$^{207}_{82}\text{Pb}$

involve alpha decays, which are the only decays that reduce the mass number. The alpha decay of a heavy nuclide increases the ratio N/Z, so some beta decays are likely to occur in the chain too.

Table 7.1 shows the four **radioactive series** by which the heaviest nuclides eventually reach stability. The alpha particle has $A = 4$, so other radioactive nuclides (with $A < 232$) lie along one of these four series. Some radioactive nuclides can experience either alpha or beta decay, so there are some alternate paths that a decay chain can follow. Eventually each series reaches the final product listed in Table 7.1.

How Are Radioactive Nuclides Used in Medicine?

Radioactive nuclides (or **radionuclides**) are used extensively in both the diagnosis and treatment of medical conditions. For diagnostic purposes, the patient either swallows or takes an injection of radionuclides. Many different radionuclides can be used, and they can be chosen with a specific target in mind. For example, to examine the thyroid gland, the iodine isotope ^{123}I is often used because iodine is readily absorbed by the thyroid. ^{123}I decays by electron capture $^{123}\text{I} + \beta^- \rightarrow {}^{123}\text{Te}$ with a half-life of 13 hours. The tellurium nuclide subsequently emits a 159-keV gamma ray, and it's the detected gamma radiation that is used to create images of the thyroid. Further, the rate of iodine absorption by the thyroid can indicate hypothyroidism or hyperthyroidism. Other radionuclides can be used to target bones, the cardiovascular system, and internal organs.

Special detection systems are used to produce three-dimensional images, which are more useful than two-dimensional images in some diagnoses. One technique, called single-photon emission computed tomography (**SPECT**), takes two-dimensional images viewed from different angles as a gamma-ray camera is rotated around the patient. The multiple images are reconstructed by computer to produce the three-dimensional image. Another three-dimensional imaging technique is positron emission tomography (**PET**), where the patient is injected with an organic solution containing a positron emitter, often ^{18}F. The emitted positrons slow down before they can travel far, and when they encounter electrons, the resulting matter–antimatter annihilation produces a pair of gamma rays ejected in opposite directions. Detection of the coincident photons produces sharp computer-generated images.

Radionuclides are used to treat diseases, such as cancer, where the emitted radiation kills or causes mutation in malignant cells. For example, ^{131}I is a beta emitter, with a half-life of 8.0 days. Like other iodine isotopes, it is readily absorbed into the thyroid gland. The emitted electrons, with kinetic energies up to 606 keV, can penetrate and damage or kill cells up to 2 mm away.

Radioisotopes are specially prepared for medical use. Normally, such isotopes are chosen to have relatively short half-lives, on the order of hours or days, so that they don't linger in a patient's system. This requires that medical radioisotopes be used fairly soon after being prepared. Radionuclides must be produced and supplied to hospitals and treatment centers daily or weekly in order to be effective. Typically, they are produced by bombardment of materials in cyclotrons or obtained as by-products from nuclear reactors. For example, the ^{123}I isotope discussed earlier is produced by bombarding xenon with high-energy protons. The ^{124}Xe isotope absorbs a proton in the reaction ^{124}Xe + ^{1}H $\rightarrow$ ^{123}Cs + 2^1n, and then the cesium decays to ^{123}I by the two successive positron emissions.

How Does Radioactive Dating Work?

Radioactive dating is a technique used to determine the age of a sample by measuring the abundance of a particular radioactive nuclide relative to other nuclides that are stable or that have much longer half-lives. One of the best known and most widely used methods of radioactive dating uses the carbon isotope ^{14}C, a beta emitter with a half-life of 5,730 years. ^{14}C is produced when nitrogen in Earth's atmosphere is struck by cosmic rays. Plants and animals then take in some ^{14}C along with the more common stable isotope ^{12}C. After an organism dies, the amount of ^{14}C left in its remains drops exponentially, while the ^{12}C remains. Therefore, the ratio of ^{14}C to ^{12}C decreases in time, and the ratio is an excellent indication of the sample's age. The half-life of ^{14}C makes this technique ideal for determining the ages of fossils going back thousands of years, but the uncertainty grows for ages greater than several half-lives.

A wider range of radioactive dating techniques is available for inorganic materials, such as old rocks. One example is the comparison of lead isotopes. The isotopes ^{206}Pb and ^{207}Pb are stable isotopes that lie at the end of the ^{238}U and ^{235}U decay chains, respectively (Table 7.1). The isotope ^{204}Pb is also stable, and no other nuclide decays to it, so it is presumed that the amount of ^{204}Pb is stable in time. For this reason, a comparison of the ratios ^{206}Pb/^{204}Pb and ^{207}Pb/^{204}Pb is a good indication of age. Newer samples have a relatively higher ratio of ^{206}Pb/^{204}Pb relative to ^{207}Pb/^{204}Pb, due to the longer half-life of ^{238}U. Due to the long half-lives of both uranium isotopes, this technique has been used to date specimens from the early Earth.

Another interesting isotope for radioactive dating is ^{36}Cl, with a half-life of 300,000 years. It is often found in groundwater and glaciers and has been used to model geological patterns up to 1 million years in the past. The isotope ^{10}Be, with a half-life of 1.36 million years, has been used for similar purposes in geology. Further, because the production of ^{10}Be depends on cosmic rays, the variation in its content in different layers of the Antarctic ice sheet has been linked to variations in solar activity.

WHAT IS FISSION?

Fission occurs when a nucleus splits into two smaller nuclei. Although some nuclides (including ^{232}Th and several of the uranium isotopes) undergo spontaneous fission, the most common form of fission is induced by a neutron striking the nucleus. The uncharged neutron is able to penetrate the nucleus to initiate the fission process. An example of a fission reaction is

$$_0^1 n + {}_{92}^{235}\text{U} \rightarrow {}_{56}^{142}\text{Ba} + {}_{36}^{92}\text{Kr} + 2\,_0^1 n \qquad (7.11)$$

Although the reaction shown in Equation (7.11) is just one of many possible ways in which ^{235}U can undergo fission, it is typical in many respects. There are normally just two nuclei (and no more) produced, with one somewhat larger than the other. It's possible for the two nuclei produced in fission to be equal or nearly equal in size, but this occurs less frequently. The release of two neutrons is typical; normally, two to four neutrons are released in each fission event. Notice, however, that the total numbers of protons and neutrons are the same before and after the reaction. That is, neither protons nor neutrons are converted into other particles by fission. Fission decay products (in this example, barium and krypton) are isotopes that are extremely neutron rich. This can be explained by the fact that the ratio of neutrons to protons, N/Z, is significantly larger for uranium than for any of the possible fission products. Thus, the barium and krypton isotopes have more neutrons than the stable isotopes of those elements have. This makes the fission products radioactive and likely β⁻ emitters. Fission always results in the production of radioactive by-products.

Not shown in the reaction in Equation (7.11) is gamma radiation, which accounts for some of the energy released in fission. The two nuclei and free neutrons all carry significant amounts of kinetic energy. The source of this energy can be understood by studying the binding energy per nucleon curve in Figure 7.9. The medium-weight isotopes that result from fission must have more binding energy per nucleon than the original uranium nucleus, so there will always be energy released in fission products. Equivalently, you can think of the released energy as coming from the loss of mass. Typically, the products of fission have total mass about 0.2 u less than the total mass of the reactants, resulting in a release of energy equal to that mass difference multiplied by c^2, which is on the order of 200 MeV (or 3×10^{-11} J) per fission event.

The naturally occurring nuclide ^{235}U is particularly susceptible to fission, as is the artificially made ^{239}Pu. Some other isotopes of uranium and thorium can fission, but this is much less likely and depends to a large degree on the energy of the incoming neutron.

What Is a Chain Reaction?

Fission normally releases between two and four free neutrons. In a sample of uranium (or other fissionable material), it's likely that one or more of the escaping neutrons will strike another nucleus to cause another fission. Each fission event releases more neutrons, so this process can continue indefinitely in what is called a **chain reaction**, shown schematically in Figure 7.10.

For the chain reaction to continue, it's required that on average one or more free neutrons from each fission event go on to generate another fission. There are several factors that affect whether this will occur in a given sample. If a sample is too small, too many neutrons will escape before striking another nucleus, and a chain reaction can't occur. Impurities in the sample can absorb neutrons and impede the chain reaction. In the case of a uranium sample, the ratio of uranium isotopes is crucial. Naturally occurring uranium consists of about 99.3% ^{238}U, which generally absorbs free neutrons rather than

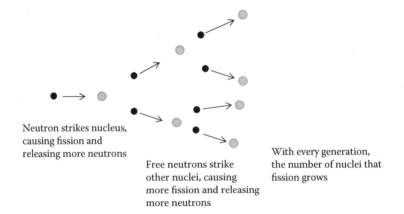

Neutron strikes nucleus, causing fission and releasing more neutrons

Free neutrons strike other nuclei, causing more fission and releasing more neutrons

With every generation, the number of nuclei that fission grows

Figure 7.10   Schematic of chain reaction.

undergoing fission. The highly fissionable ^{235}U isotope makes up only 0.7% of natural uranium. There are several methods of "enriching" uranium to more than 0.7% ^{235}U before starting the chain reaction, such as running a uranium-containing compound through a centrifuge.

A chain reaction can proceed very quickly, due to the close proximity of atoms in a sample. In order for the chain reaction to be controlled—as it is in an experimental setting or nuclear reactor—the average number of neutrons per fission that generate a secondary fission must be very close to one. If this number is even slightly larger—say, 1.1—the result is an exponential growth in the rate of fission and an uncontrolled chain reaction.

What Makes an Atomic Bomb Work?

An atomic bomb is the result of an uncontrolled chain reaction. Making an atomic bomb requires a significant sample of fissionable material (on the order of 10 kg or more), either enriched uranium or plutonium. The fissionable material is assembled quickly in the vicinity of a neutron source that initiates the chain reaction.

How Much Energy Is Released in Fission?

The energy involved in fission reactions—and in any reaction involving the nucleus—is several orders of magnitude greater than in chemical reactions. Earlier we noted that 3×10^{-11} J is a typical amount of energy released in a fission event. That translates to about 8×10^{10} J (or 80 GJ) per gram of fissionable nuclide, normally ^{235}U or ^{239}Pu. In the case of natural uranium, which contains only 0.72% ^{235}U, that's still over 5×10^8 J per gram of natural uranium, assuming that all the ^{235}U nuclei undergo fission and the ^{238}U nuclei do not.

By comparison, chemical reactions used to supply most energy needs produce much less energy. For example, burning coal generates about 3×10^4 J (30 kJ) per gram. For petroleum and natural gas (methane), the results are similar: about 40 kJ/g and 50 kJ/g, respectively. The much higher energy content for fissionable material makes fission reactors viable for large-scale energy production. The same difference in energy content is what makes nuclear weapons so destructive and terrifying. The energy released in one of the two atomic bombs dropped on Japan in 1945 was roughly equal to the energy contained in all the bombs dropped on England by Germany in the 1940–1941 "blitz" campaign.

How Does a Nuclear Reactor Work?

A nuclear (fission) **reactor** is a device that contains fissionable material and is designed to sustain a chain reaction at a steady rate. Most reactors are designed primarily to produce electrical energy on a large scale. Reactors can also be used to make fuel for nuclear weapons and radioactive isotopes for medical, industrial, or research purposes.

The first reactors were constructed in the United States in World War II, originally to prove the concept of the chain reaction and to generate scientific data about the nature of fission. In the last year of the war, large reactors at Hanford, Washington, produced plutonium, a highly fissionable material used to make atomic bombs. In a reactor fueled by uranium, fission takes place principally in the ^{235}U isotope. The ^{238}U isotope absorbs neutrons to form ^{239}U, which subsequently undergoes two beta decays:

$$^{239}_{92}U \rightarrow {}^{239}_{93}Np + {}^{0}_{-1}\beta^- + \overline{\nu} \qquad t_{1/2} = 23 \text{ min}$$

$$^{239}_{93}Np \rightarrow {}^{239}_{94}Pu + {}^{0}_{-1}\beta^- + \overline{\nu} \qquad t_{1/2} = 2.4 \text{ d}$$

^{239}Pu undergoes alpha decay, but with a half-life of 24,000 years, it is essentially stable in the reactor and can be separated from the other fission products and unused uranium fuel.

Modern reactors are designed for the large-scale production of electrical energy. Commercial reactors typically use uranium fuel that has been enriched to 4%–5% ^{235}U. Even with this enrichment, it's necessary that the reactor contain a **moderator** material, which slows down the emitted neutrons to keep too many of them from being absorbed by ^{238}U nuclei. Water, graphite, and beryllium are all good moderators. It's essential for safety that the reactor contains **control rods,** made of a good neutron-absorbing material such as cadmium, which can be inserted quickly between the uranium fuel cells to shut down the chain reaction in case of emergency.

One common type of commercial reactor is the boiling water reactor (**BWR**), shown schematically in Figure 7.11. Heat from the reactor core is used to boil water (which is also used as the moderator), and the steam is used to generate electricity in a turbine. In a pressurized water reactor (**PWR,** Figure 7.12), energy from the fission reaction heats and pressurizes water, which is then circulated through a heat exchanger to generate steam. The main difference between the two types of reactors is that, in the BWR, steam goes directly from the reactor core to the turbine, while, in the PWR, the steam driving the turbine is one step removed from the core. Typical power plants (of both varieties) can produce electrical energy at a rate on the order of 1000 MW.

As of 2012 there were 104 working commercial power reactors in the United States (35 BWR and 69 PWR), which produced about 19% of the nation's electricity. Worldwide, about 14% of electricity is generated by nuclear plants, but the distribution varies widely. Many countries have no nuclear capacity, but some countries get a much larger share of their electricity from nuclear plants, including France (78%), Belgium (54%), and Ukraine (47%).

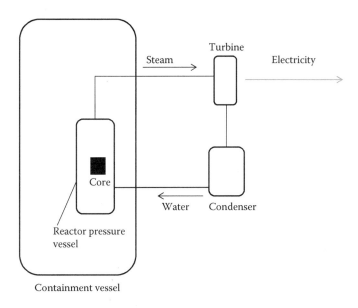

Figure 7.11   Schematic of a boiling water reactor.

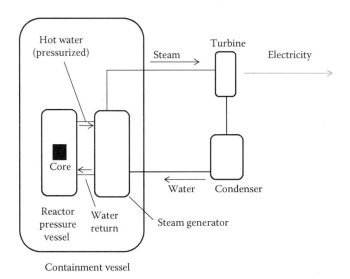

Figure 7.12   Schematic of a pressurized water reactor.

One big advantage of nuclear power plants is that they emit no fossil-fuel pollution. However, they do emit thermal pollution, raising the temperature of the local water supply. The fission fragments and spent fuel constitute radioactive waste, which must be stored securely and not released into the environment. There have been several well-known reactor accidents. In 1979 there was a relatively minor but well-publicized accident at Three-Mile Island in Pennsylvania. Accidents at Chernobyl, Ukraine, in 1986 and Fukushima Daiichi, Japan, in 2011 were much more serious, with each releasing large amounts of radiation into the surrounding environment. Such incidents raise serious questions about the long-term practicality of nuclear energy, although it is clear that many countries will continue to depend on nuclear reactors for a large share of their energy production.

WHAT IS FUSION?

Fusion is the joining of lighter nuclei to form heavier nuclei. Analogous to the release of energy in fission, the binding energy per nucleon curve (Figure 7.9) explains this energy release because the heavier products that result from fusion have more binding energy per nucleon than the initial lighter nuclei. Also analogous to fission, you can understand the energy release as a conversion of mass into energy. One example is the fusion of deuterium (^{2}H) and tritium (^{3}H),

$$^2H + {}^3H \rightarrow {}^4He + {}^1n$$

in which the mass of the products is less than the mass of the reactants by about 0.019 u. The energy released is this "lost" mass multiplied by c^2, which comes to 17.6 MeV. At first glance this might appear to be less efficient than a fission reaction, which normally releases about 200 MeV. However, because only five nucleons were involved, the energy released per nucleon is more than 3 MeV, compared with less than 1 MeV/nucleon for fission.

There are few places where fusion can occur naturally, due to the strong electrical repulsion between two positively charged nuclei at short distances. In the cores of stars (including our sun), the high temperature creates a plasma, in which the electrons are not bound to the nuclei. Then the extremely high pressure and temperature allow the hydrogen nuclei (protons) to fuse into helium. Several steps in the reaction are as follows:

$$^1H + {}^1H \rightarrow {}^2H + \beta^+ + \nu$$

$$^1H + {}^2H \rightarrow {}^3He + \gamma$$

$$^3He + {}^3He \rightarrow {}^4He + {}^1H + {}^1H$$

Fusion of hydrogen into helium is the source of the sun's energy. The sun is about 4.5 billion years old, which is roughly half of its expected lifetime. Its current composition is about three-fourths hydrogen and one-fourth helium, with less than 2% composed of heavier elements, mostly oxygen and carbon. About 5 billion years from now, the sun will expand into a red giant phase, enveloping Earth, and later the sun will evolve into planetary nebula and white dwarf stages (Chapter 10).

In the later stages of the lifetime of some stars, helium can fuse to form carbon, oxygen, and even heavier elements. Fusion can normally proceed only to ^{56}Fe, at the peak of the binding energy per nucleon curve (Figure 7.9). Heavier elements are made in the extreme conditions of a supernova event.

What Is a Hydrogen Bomb?

Fusion energy is exploited in hydrogen (or thermonuclear) bombs. The extreme conditions needed to initiate fusion require that the device contain a fission bomb as a "trigger" to start fusion in some combination of deuterium, tritium, and lithium. While the ultimate size and energy content of a fission bomb are limited by the pace of the chain reaction, there is no such limit in a hydrogen bomb. After triggering, the fusion of the bomb's light elements can occur almost instantaneously. The destructive output of hydrogen bombs reaches many **megatons**—millions of tons of TNT equivalent—compared with the 20-kiloton equivalent fission bombs used in World War II.

Hydrogen bombs were first developed by the United States and Soviet Union in the 1950s. This led to an arms race, with both countries stockpiling large numbers of weapons. The United States and Russia still have significant stockpiles, and several other countries have smaller numbers of hydrogen bombs.

What Are the Prospects for Large-Scale Energy Production from Fusion?

The high efficiency of the fusion reaction suggests that fusion could be a good source of electrical energy. Compared with fission energy, there are some obvious advantages. First, there is a plentiful supply of hydrogen in the world's water supply, compared with the relatively scarce supply of uranium used to fuel fission reactors. Second, most fusion by-products (helium and lithium isotopes) are not radioactive and do not pose the threat of fission by-products, although the free neutrons released in fusion do pose some risk. Compared with fossil-fuel energy sources, which now provide the bulk of the energy used in the United States and worldwide, the supply of fuel for fusion is almost limitless, and fusion would generate no greenhouse gases or particle pollutants.

Unfortunately, it is difficult to duplicate on Earth the conditions of extreme pressure and temperature that generate fusion in a star's core. In order for fusion energy to be practical, it's necessary that the energy produced be greater than the energy input used to initiate fusion. There are two distinct approaches scientists have used to create fusion in the lab. **Magnetic confinement** uses a toroidal magnet to trap the plasma, which can then be heated to initiate fusion. **Inertial confinement** uses a solid pellet of hydrogen (or hydrogen isotopes deuterium and tritium), with fusion initiated by multiple high-powered laser strikes on the pellet.

Scientists measure the efficiency of the fusion reaction using the parameter Q, defined as the ratio of the power output to power input. For "breakeven" conditions, $Q = 1$ is required, and this has been achieved experimentally. However, for practical energy production, a much larger Q is needed because not all of the energy released in fusion can be harnessed and transformed into electrical energy. There is currently no way to use fusion as a net supplier of energy on a commercial scale. However, there is ongoing work to improve both inertial and magnetic confinement methods. One of the most promising efforts is the internationally supported **ITER** (formerly the International Thermonuclear Experimental Reactor) program, which proposes to build, by 2020, a magnetic-confinement reactor that is designed to output 500 MW while using only 1/10 of that amount of power as input. Researchers hope that this work will be successful and can be scaled up to commercial operation. Can you imagine a world with almost limitless clean energy? Fusion may be the way to get there.

FURTHER READINGS

Krane, Kenneth S. 1987. *Introductory Nuclear Physics.* New York: John Wiley & Sons.
Lewis, E. E. 2008. *Fundamentals of Nuclear Reactor Physics.* Amsterdam: Elsevier/Academic Press.
Thornton, Stephen T., and Rex, Andrew 2013. *Modern Physics for Scientists and Engineers,* 4th ed. Boston: Cengage Learning.

Chapter 8

Fundamental Particles and Forces

Some of the most exciting recent discoveries in physics have to do with fundamental particles. For many centuries people have observed matter and wondered about its composition. A common working assumption, which goes back as far as the ancient Greeks, is that there exists a relatively small number of elementary or fundamental particles, and that the vast variety of things we see in nature can be explained by different arrangements of those few particles. Starting in the late nineteenth century, physicists began to develop the kinds of experimental tools that can determine the ultimate composition of matter. Throughout the twentieth century, more refined and elaborate equipment revealed many details about the particles that make up everything. Patterns began to emerge and, based on those patterns, theories were developed to help physicists classify and understand the particles' properties and their interactions with one another. Along with this came a new conception of fundamental forces as the manifestations of particle exchanges. The connection between fundamental particles and forces is one of the most important discoveries in modern physics.

 ## WHAT ARE FUNDAMENTAL AND COMPOSITE PARTICLES?

The idea of a particle may seem obvious enough. You might consider as an example the smallest thing you can see, such as a grain of dust or pollen. However, those objects are very large compared to what a physicist thinks of as a particle. Your vision is limited by the focusing power of your eyes, and even when aided by a good microscope, you can't see detail finer in size than the wavelength of the light you're using to view the object. Atoms are much smaller than a grain that you can see, and fundamental particles are much, much smaller than atoms.

A **fundamental particle** is one that can't be broken down into anything smaller. A **composite particle** is made up of multiple fundamental particles. There are relatively few fundamental particles, and their properties are fairly well understood today. Fundamental particles join to form neutrons and protons (and other composite particles), which in turn join to form the nuclei of all the different atoms. Atoms in turn join to form an even larger number of different molecules, which also combine or interact with one another. Indeed, there's quite a long chain linking fundamental particles with the smallest particles you can see!

WHAT ARE THE HISTORICAL ROOTS OF THE SEARCH FOR FUNDAMENTAL PARTICLES?

The idea of a fundamental particle was conceived in ancient Greece by Democritus and Leucippus (Chapter 7). They called the fundamental particle an atom, meaning literally something that can't be cut. Today the things we call atoms are composite, but we have found different, smaller fundamental particles.

The search for the composition of matter led to the first discovery of a fundamental particle, the electron, by J. J. Thomson in 1897 (Chapter 7). Along with the negatively charged electron, the other piece of a hydrogen atom is the positively charged proton. The neutron, a neutral particle with mass just slightly larger than the proton's, was discovered by James Chadwick in 1932. Protons and neutrons combine to form the nuclei of all atoms, and an atom consists of a nucleus and electron(s). Thus, all atoms are composed of protons and neutrons (which we now know are both composite, not fundamental) in a nucleus, along with electrons (which are fundamental).

WHAT IS ANTIMATTER?

Protons, neutrons, and electrons are examples of ordinary matter. In 1932, less than a year after the discovery of the neutron, American physicist Carl Anderson discovered the **positron,** a form of **antimatter.** Anderson was studying the effects of cosmic rays, which are high-energy particles that reach Earth from sources mostly outside our solar system. Anderson noticed that when cosmic rays interact with matter, they sometimes produce particles with the same mass as an electron but opposite charge (i.e., a charge $+e$, compared with the electron's $-e$). Anderson originally referred to the new particle as a "positive electron," but today we call it a positron.

The positron was the first known **antiparticle,** or particle of antimatter. An antiparticle characteristically has the same mass and spin but opposite charge of its corresponding particle. The British physicist Paul A. M. Dirac had pre-dicted the existence of antimatter on theoretical grounds, several years before Anderson's discovery. There are many more examples of antiparticles. For example, the **antiproton** has the same mass as the proton but opposite charge (−e, compared with the proton's +e).

What Is the Relative Abundance of Matter and Antimatter in the Universe?

A particle and its antiparticle exhibit the same physical properties with the exception of their electric charges, which are opposite to one another. Based on this symmetry, you might expect matter and antimatter to be equally abundant in our universe. However, that's not the case. Ordinary matter dominates everywhere in the universe, as far as we can tell. Why there is such an imbalance of matter over antimatter is currently an open question, and physicists consider it one of the more interesting questions they're trying to answer.

Matter and antimatter cannot coexist in the same space because any attempted combination of the two results in annihilation of both particles. In such a process the two particles' mass–energy is converted to a pair of gamma-ray photons. The energy produced in any matter–antimatter annihilation is significant because it comes from the complete conversion of mass to energy via Einstein's $E = mc^2$ mass–energy relation. For example, an electron or positron at rest has a mass–energy of 0.51 MeV. When they annihilate, each of the two gamma rays (traveling in opposite directions, as required by momentum conservation) has energy 0.51 MeV. Annihilation of more massive particles (such as a proton–antiproton pair) produces more energy.

WHAT IS THE PARTICLE ZOO?

Because we don't see antimatter mixed with ordinary matter, antimatter is observed only in certain natural processes (such as when Anderson found the positron) or else is created in the laboratory. One of the principal reasons why physicists have developed larger and larger particle accelerators over the last century is to create new particles. This is the reverse of the particle–antiparticle annihilation process because, in this case, the $E = mc^2$ relation relates the kinetic energy of accelerated particles to the new mass they create when they give up that kinetic energy in a collision.

For example, the antiproton was discovered in 1955 by a team of scientists working at the University of California, Berkeley. They used a newly commissioned **cyclotron,** which was dubbed the bevatron because it accelerated charged particles to energies of billions of electron volts (or GeV, where

1 GeV = 10^9 eV = 1 billion eV). Such high energies are needed to produce antiprotons, which have a rest energy of 0.938 GeV. A high-energy proton that strikes a proton in a target can initiate the reaction

$$p + p \rightarrow p + p + p + \bar{p} \tag{8.1}$$

where p is the proton and $\bar{p}$ the antiproton. (Antiparticles are normally designated by placing the over-bar above the corresponding particle symbol.) Note that a third proton is produced along with the antiproton, as required by charge conservation and other conservation laws.

GOING DEEPER—CYCLOTRONS AND SYNCHROTRONS

The cyclotron was a device invented in the 1930s by American physicist Ernest O. Lawrence for the purpose of accelerating charged particles to high energies. Lawrence knew that the best way to study the structure of matter was to bombard it with high-energy particles, and he created the cyclotron for this purpose.

The cyclotron has a flat, circular shape. A magnetic field perpendicular to the plane of the circle creates a force on any charged particle moving in the plane. A particle with charge q and momentum p will move in a circular orbit of radius $R = p/qB$ in the magnetic field B. The circular region is divided into two halves called "dees" because half of the circle resembles the shape of the letter D. Between the dees there is a voltage that alternates back and forth at the cyclotron frequency

$$f = \frac{qB}{2\pi m} \tag{8.2}$$

for a particle of mass m. Each time the particle jumps from one dee to another, the voltage gives it a push forward to accelerate it, and the alternating voltage ensures that the particle increases its speed every time it jumps the gap. Because the radius of curvature is proportional to the particle's momentum, the particle begins in the middle and spirals out to the cyclotron's outer radius, where it achieves maximum energy. Lawrence found that he could accelerate protons to kinetic energies of over 100 MeV in a large cyclotron (nearly 5 m in diameter).

Because the highest particle energy is limited by the cyclotron's diameter, a different particle accelerator design, called a **synchrotron,** is now used in high-energy experiments. The synchrotron consists of a ring of fixed radius. As a charged particle accelerates around the ring, the magnetic field is increased to keep the path radius fixed while the particle's speed increases. The ring can be made very large because, unlike a cyclotron, the magnetic field of the synchrotron needs to be present only along the particle's path, not the entire plane of the circle. Using the synchrotron design, particle energies of 900 GeV (= 9×10^{11} eV) were achieved at the Fermilab accelerator in Illinois in 1986, with a ring radius of 1.09 km. This accelerator was used to confirm the existence of the top quark in 1995. The larger (4.3 km radius) synchrotron at Europe's CERN facility reached 4 TeV energy (= 4×10^{12} eV) by 2012 and in the same year produced evidence of the Higgs boson.

The synchrotron device at CERN has been repurposed and renamed the **Large Hadron Collider (LHC)** because it's designed to create and study collisions of high-energy protons traveling in opposite directions. Although difficult to achieve, the colliding-beam technology is a great advance over older particle accelerators, in which a beam of high-energy particles was fired into a fixed target. When an accelerated particle strikes a fixed target, much of the energy is wasted due to recoil of final products, as required by momentum conservation. However, when identical particles having the same energy collide head-on, all their energy is available to react and produce new particles. Thus, in a head-on collision between two 4 TeV protons, a total of 8 TeV is available.

After the positron was discovered in 1932, a number of other particle discoveries followed. In 1937 physicists found the **muon,** a negatively charged particle that appeared similar to an electron but more than 200 times heavier. The **pi meson** (or **pion**) and **K-meson** (or **kaon**) were both found in 1947. Mesons are named from the Greek root meso- (middle) because their masses are intermediate between electrons and protons. Both pions and kaons come in positive, negative, and neutral forms. In 1956 another important class of particles, the **neutrino,** was first observed. Neutrinos are neutral and very small—so small that they were at first thought to be massless. Neutrinos come in six different types, but all are neutral. The sixth neutrino was not confirmed until 2000.

By the 1960s and 1970s, the number of known particles was large and growing. Although some particles exhibited similarities with one another (for example, electrons and muons), many (for example, protons and neutrinos) appeared quite different, and the vast collection of particles became known as the **particle zoo.** Today the particles and how they are related to one another are much better understood, using a fairly simple set of classifications. The zoo has been tamed.

HOW DO PARTICLES MEDIATE FORCES?

An important insight that helped physicists understand particles came from Japanese physicist Hideki Yukawa in the 1930s. Yukawa sought to explain the strong force (or nuclear force)—the attractive force between protons and neutrons that binds them together to form nuclei (Chapter 7). Yukawa suggested that some new particle, which had not been observed at that time, was exchanged between neighboring protons and neutrons, resulting in a net attractive force between them. He predicted that the new particle would have a mass of about 100 MeV/c^2. That's heavier than an electron (0.51 MeV/c^2) but lighter than a proton (938 MeV/c^2), so it was called a meson. Yukawa's prediction was conformed with the discovery of the pion in 1947. (The several pions have masses between 135 and 140 MeV/c^2.) The pion is said to be the mediator or force carrier of the nuclear force. Similarly, the electromagnetic force can be attributed to the exchange of photons between charged particles. Physicists believe that mediating particles are responsible for all the fundamental forces.

 WHAT ARE QUARKS?

In the early 1960s, physicists Murray Gell-Mann and George Zweig suggested how many of the known particles might be understood as composed of smaller, fundamental particles. Gell-Mann called the new fundamental particles **quarks,** after a line from James Joyce's novel *Finnegans Wake:* "Three quarks for Muster Mark." It appeared then that three different quarks would be necessary to explain the compositions of known particles, and the quarks were named **up, down,** and **strange.** The strange quark was so named because some unusual behavior was observed in the decay of kaons, which contain this kind of quark. Protons and neutrons are composed of different combinations of up and down quarks, and therefore so are all the nuclei in ordinary matter.

Physicists now recognize six kinds of quarks, which are listed in Table 8.1. The properties of the quarks help explain the properties of all the composite particles that contain quarks, which together are called **hadrons.** In particular, the fractional charges of the quarks (+2e/3, +e/3, −e/3, or −2e/3) explain why most observed hadrons have a charge of zero or ±e. Quarks are never found alone but rather come in combinations of two or three. Any combination of three quarks forms a type of particle called a **baryon** (from the Greek root for heavy). Protons and neutrons are examples of baryons. A proton contains two up quarks and one down, giving a net charge of +2e/3 +2e/3 −e/3 = +e. A shorthand designation uses "u" for up and "d" for down, with the proton's quark composition given as uud. Similarly, a neutron has quark composition udd, which makes its net charge zero, as observed. Like the baryons themselves, quarks are

TABLE 8.1   THE SIX QUARKS

Quark Name	Quark Symbol	Electric Charge	Approximate Mass (GeV/c^2)
Up	u	$2e/3$	0.002
Down	d	$-e/3$	0.005
Strange	s	$-e/3$	0.09
Charmed	c	$2e/3$	1.25
Bottom	b	$-e/3$	4.2
Top	t	$2e/3$	172

TABLE 8.2   QUARK COMPOSITIONS OF SELECTED BARYONS AND MESONS

Baryon	Quark Composition	Meson	Quark Composition
p (proton)	uud	π^+	$u\bar{d}$
n (neutron)	udd	π^-	$\bar{u}d$
Σ^+	uus	D^+	$c\bar{d}$
Σ^0	uds	D^0	$c\bar{u}$
Ω^-	sss	K^+	$u\bar{s}$
Λ	uds	K^0	$d\bar{s}$
Λ_C^+	udc	J/ψ	$c\bar{c}$
Ξ^0	uss	T^0	$t\bar{u}$
Ξ^-	dss	B^-	$\bar{b}u$

subject to the strong force through exchanges of particles, which explains how they bind together to form baryons and mesons.

The quark compositions of some selected baryons are shown in Table 8.2. Each baryon has an antiparticle, composed of three antiquarks. For example, the antiproton's quark composition is $\overline{uud}$, two up antiquarks and a down antiquark. Each of the six quarks listed in Table 8.1 has an antiquark with opposite charge. Thus, in the antiproton, the net charge is $-2e/3 -2e/3 +e/3 = -e$ as expected.

Because free quarks are not observed, quark masses are difficult to determine, and the masses listed in Table 8.1 are only approximate. Further, the mass of a baryon or meson is not simply the mass of its quark constituents. That's because the quarks are in such close proximity that their binding energy contributes significantly to the observed composite particle mass, and that mass depends strongly on the arrangement of quarks within the particle. For example, a proton composed of two up quarks and one down quark has a mass of 0.938 GeV/c^2, but the sum of the three quark masses is only about 0.01 GeV/c^2. In contrast, a pion, made up of an up quark and a down antiquark, has a mass of 0.14 GeV/c^2—still much greater than the net quark mass but much lighter than the proton.

Table 8.2 also shows that mesons are composed of a quark–antiquark pair. With fewer quarks they tend to be lighter than baryons. As with baryons, different quark combinations yield mesons having net charge zero or ±e. For example, the positive pion π^+ has quark composition $u\bar{d}$ and a net charge +2e/3 +e/3 = +e. The neutral kaon K^0 is $d\bar{s}$ and has net charge −e/3 +e/3 = 0.

What Is Color Charge?

Color charge is another property of quarks. The name can be misleading because it has nothing to do with colors we see (which would be meaningless on such a small, subparticle level) and nothing to do with electric charge, which is already accounted for by each quark's fractional charge (Table 8.1). It's better to think of color charge as a separate quantum number independent of quantum numbers associated with energy level, angular momentum, and other measurable quantities. The theory of color charge is called **quantum chromodynamics.**

The reason color charge is needed is that quarks are spin-1/2 particles, making them fermions. Therefore, they obey the Pauli exclusion principle (Chapter 4), which means that no two of them can be in the same quantum state. Without an additional quantum number, the two up quarks in a proton would violate the Pauli principle. It's possible to have a baryon made up of three of the same kind of quark, such as the Ω^- (sss) and the Δ^{++} (uuu). Therefore, three kinds of color charge are needed so that each quark in a baryon can have a different color. The names chosen as colors are red, blue, and green (designated R, B, and G). Colors associated with antiquarks are different and are designated antired, antiblue, and antigreen ($\bar{R}, \bar{B}, \bar{G}$).

Although color charge has nothing to do with the colors we see, the choices of red, blue, and green are not arbitrary. In a baryon containing quarks of all three colors, the net effect is no color or "white," analogous to the way spectral colors combine to form white in our perception. This is a valuable analogy in particle physics because the color charge does not affect how two baryons interact with one another. In effect, each one sees the other as colorless.

What Are Gluons?

The **gluon** is the particle that mediates the strong force between quarks. Although gluons have not been observed, they must exist because of the strong binding between quarks in baryons and mesons. The gluon is a massless particle, and it works by having a color–anticolor pair, such as $R\overline{G}$. Two quarks of different color attract each other through the interaction of an appropriately colored gluon. For example, a red quark and a blue quark interact through the exchange of a $B\overline{R}$ gluon. Because of the gluon exchange, each quark changes color in the process.

Gluons are the ultimate force carrier for the strong force that acts between quarks or between hadrons, which are composed of quarks. Earlier in this chapter, the force interaction between hadrons was modeled using the pion as the mediating particle. This works fine at relatively low energies, and in this case the net force between hadrons is called the residual strong force. The more complete theory of the strong force, which includes the forces between quarks within hadrons, requires gluons.

What Is Quark Confinement?

Free quarks have not been observed, and physicists believe they never will be. Rather, quarks can only exist bound to one another within hadrons, a concept known as **quark confinement.** Confinement works because, unlike the gravitational or electromagnetic forces, which decrease in strength with increasing distance, the gluon-mediated strong force actually increases when quarks are pulled further apart. Any attempt to separate the quarks by supplying more energy (for example, in a particle accelerator) will not free the quarks. At sufficiently high energies, the excess energy leads to the creation of new particles, as in Equation (8.1). This is analogous to trying to create a magnetic monopole by separating the two parts of a magnetic dipole, which only leads to the creation of new dipoles (Chapter 2).

WHAT ARE LEPTONS?

Quarks are fundamental particles. Another type of fundamental particle is the **lepton.** Unlike quarks, leptons do exist independently of one another and are not affected by the strong force. There are three **generations** (or families) of leptons, with each generation consisting of a negatively charged particle and a corresponding neutrino. Each of the six leptons has an antiparticle, so taking these into account, the total number of leptons is 12. The three leptons (electron, muon, and tau) are negatively charged, so their antiparticles are

TABLE 8.3   LEPTON PROPERTIES

Particle	Symbol	Antiparticle	Mass (MeV/c^2)
Electron	e^-	e^+	0.511
Electron neutrino	ν_e	$\bar{\nu}_e$	$<2.2 \times 10^{-6}$
Muon	μ^-	μ^+	105.7
Muon neutrino	ν_μ	$\bar{\nu}_\mu$	<0.17
Tau	τ^-	τ^+	1776.8
Tau neutrino	ν_τ	$\bar{\nu}_\tau$	<15.5

positively charged. The antineutrinos, like the neutrinos themselves, have zero charge. Some of the properties of the leptons and their antiparticles are listed in Table 8.3.

The electron is the smallest lepton (excluding the neutrinos) and is found in ordinary matter. It was the first fundamental particle to be isolated, when J. J. Thomson measured its charge-to-mass ratio in 1897. The muon was found in 1937, and the third lepton, the massive **tau** particle, was confirmed in 1977. The electron is stable—meaning that it doesn't decay into any other particles and has an essentially infinite lifetime. A muon decays into an electron and neutrinos with a half-life of just 2.2×10^{-6} s. The tau has an even shorter half-life, on the order of 10^{-13} s, and can decay into a muon (and neutrinos) or an electron (and neutrinos). Unlike the electron and muon, the tau is massive enough that it can also decay into hadrons.

Long before the neutrino was found in the lab, there were hints of its existence from experiments involving beta decay of radioactive nuclei. When a nucleus undergoes beta decay, it emits an electron spontaneously. In the process, the nucleus loses some of its mass–energy, and that energy shows up as kinetic energy in the recoiling pieces. However, the emitted electron and recoiling nucleus do not always have the maximum allowed kinetic energies. This is unlike alpha decay, where all the lost mass–energy is accounted for in the recoiling nucleus and alpha particle. In 1930 Austrian physicist Wolfgang Pauli suggested that a small neutral particle, at that time not yet observed, was responsible for carrying away the remainder of the kinetic energy in beta decay. It would have to be neutral because charge is conserved without taking it into account, and the dynamics of the beta decay reaction suggested an extremely small, perhaps even massless particle. The name neutrino (meaning small, neutral particle in Italian) was coined by Enrico Fermi, who supplied a complete theory of beta decay in 1934.

In the form of beta decay that produces an electron, it's actually an electron antineutrino $\bar{\nu}_e$ that's produced. Decay of the ^{14}C isotope (a typical beta decay) looks like this:

$$^{14}C \rightarrow {}^{14}N + e^- + \bar{\nu}_e \qquad (8.3)$$

In the form of beta decay that produces positrons, the positron is accompanied by an electron neutrino ν_e. For example, a common isotope used in the medical procedure positron emission tomography (PET) is ^{23}Mg, which decays via

$$^{23}Mg \rightarrow {}^{23}Na + e^+ + \nu_e \qquad (8.4)$$

Because neutrinos have zero electric charge and are very small, they are difficult to detect. As an indication of how weakly neutrinos interact with matter, it's estimated that the flux of solar neutrinos reaching Earth is perhaps 60 billion per square centimeter per second. They pass virtually without interaction through Earth—and you! The first successful neutrino detection came in 1956 in an experiment designed by Clyde Cowan and Frederick Reines. They reacted antineutrinos produced in beta decay with protons, a reaction that yields a neutron and a positron. The positron then meets an electron, and the gamma rays resulting from the electron–positron annihilation are detected, signaling the antineutrino's presence from the start.

It's now believed that the neutrinos have mass, albeit very small. Difficulty in detection makes it impossible to estimate the masses precisely, so upper limits are shown in Table 8.1. The electron neutrino has less than 1/100,000 times the mass of its corresponding lepton, the electron.

What Is Neutrino Oscillation?

After the existence of neutrinos was confirmed, physicists worked to build new and larger neutrino detectors. A successful scheme used for many years involved using large tanks filled with the cleaning fluid C_2Cl_4. Incident neutrinos—mostly those arriving from the sun—initiated the reaction

$$\nu_e + {}^{37}Cl \rightarrow {}^{37}Ar + e^- \qquad (8.5)$$

with the subsequent decay of the radioactive isotope ^{37}Ar used to signal the presence of the neutrino. Experiments over many years found a regular stream of neutrinos incident from the sun but at a consistently lower rate than expected by theory.

This so-called "solar neutrino problem" led to further study using a new kind of detector filled with ultrapure water, which is rich in protons (hydrogen nuclei). An antineutrino creates the reaction

$$\bar{\nu}_e + {}^1\text{H} \rightarrow {}^1 n + e^+ \tag{8.6}$$

The resulting positron generates light, which is detected by thousands of photomultiplier tubes set up around the outer shell of the detector. Such detectors in Sudbury, Ontario, and Kamioka, Japan, each found evidence that neutrinos change spontaneously from one kind to another, a process called **neutrino oscillation.** The fact that some solar neutrinos mutated from electron neutrinos to some other form during the time it takes them to travel to Earth explained the solar neutrino problem. Neutrino oscillation is also related to the fact that neutrinos have mass, which seems to be confirmed by other experiments. Physicists continue to study neutrino oscillation, not only for what it tells us about neutrinos but because it's related to other interesting problems, such as the amount of matter in the universe and the dominance of ordinary matter over antimatter.

 ## WHAT IS THE WEAK FORCE?

While the strong force affects only hadrons, the **weak force** (or **weak interaction**) affects leptons and, in fact, all fermions, particles with half-integer spin. It's one of the four fundamental forces in nature along with gravity, electromagnetism, and the strong force. In the 1960s a successful theory was developed that combined the weak force and electromagnetism into a single **electroweak interaction.**

The weak force is mediated by three particles called W^+, W^-, and Z^0, which are all bosons—particles with integer spin. They are all massive particles, with masses ranging from 80 GeV/c^2 to 91 GeV/c^2, and they all have lifetimes shorter than 10^{-24} s. After an intense search, the W^+, W^-, and Z^0 were all detected in 1983.

Specifically, the weak interaction governs the process of beta decay. For example, in the form of beta decay resulting in emission of an electron, a neutron with quark configuration udd effectively changes into a proton with quark configuration uud. Thus, on the level of quarks, one down quark has changed into an up. This is accomplished through emission of a W^- particle (note the charge conservation in this process), which quickly decays into the electron and electron antineutrino.

 ## WHAT IS A FEYNMAN DIAGRAM?

Feynman diagrams (named for twentieth century American physicist Richard Feynman) are used to illustrate processes involving interactions between particles, including the mediating particles associated with forces. An example of a Feynman diagram is Figure 8.1, which illustrates the electromagnetic force between a pair of electrons. In this diagram, time is represented on the vertical axis and space on the horizontal axis. The electrons approaching one another interact through exchange of a gamma-ray photon. Figure 8.2 shows a proton and neutron interacting by the nuclear force through the exchange of a neutral pion, and Figure 8.3 illustrates the role of the W⁻ particle in beta decay.

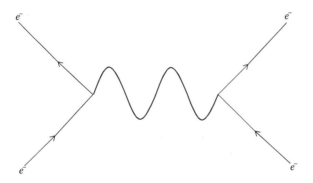

Figure 8.1   Feynman diagram illustrating the electromagnetic interaction between electrons. The wavy line represents the photon, which mediates the interaction.

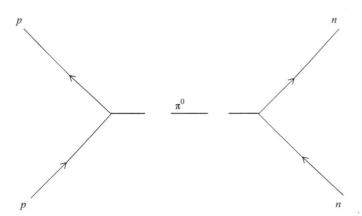

Figure 8.2   Feynman diagram illustrating the strong nuclear interaction between a proton and neutron, mediated by the π^0.

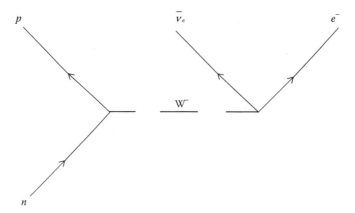

Figure 8.3 Feynman diagram showing the process of beta decay, mediated by the W⁻, resulting in the emission of an electron and an antineutrino.

WHAT CONSERVATION LAWS ARE ASSOCIATED WITH HADRONS AND LEPTONS?

Conservation laws are central to the study of physics. Physicists believe that electric charge is conserved without exception in nature. For example, in experiments that create antiprotons (Equation 8.1), another positive charge has to be created along with the negative antiproton so that the net charge is the same after the reaction as before. Mass–energy is another important conserved quantity. Mass and energy may be exchanged with one another, following the relationship $E = mc^2$, but the net mass–energy does not change. A subtle example of this is binding energy. The mass of a bound nucleus is less than the sum of the masses of all the protons and neutrons in the nucleus by an amount exactly equal to the binding energy of the nucleus.

There are other conserved quantities that apply to hadrons and leptons. All baryons are assigned a baryon number $B = 1$ (or $B = -1$ for antiparticles). Other particles (mesons and leptons) have $B = 0$, and baryon number is conserved in reactions. For example, in beta decay a neutron with $B = 1$ becomes a proton, which also has $B = 1$. The electron and antineutrino have $B = 0$, so baryon number is conserved. The stability of the proton is linked to baryon conservation. The proton is the lightest baryon, so there are no smaller baryons into which it can decay. Therefore, proton decay (which has not been observed) would require violation of baryon conservation. Individual quarks are assigned baryon number 1/3, which explains why baryons (composed of three quarks) have $B = 1$ and mesons (a quark–antiquark pair) have $B = 1/3 + (-1/3) = 0$.

Another property, called **strangeness,** is the property of strange quarks. Strangeness is quantified as –1 for strange quarks and +1 for the strange anti-quark. Strangeness is conserved in the electromagnetic and strong interactions but not in the weak interaction. Thus, a positive kaon K⁺ with quark configuration $u\bar{s}$ can decay into two positive pions ($u\bar{d}$) and one negative pion ($\bar{u}d$) via the weak interaction. Each of the charmed, bottom, and top quarks has its own conserved quantities, analogous to strangeness.

Each of the three families of leptons—electron, muon, and tau—has a conservation number associated with it. For example, the electron lepton number is $L_e = 1$ for the electron and its neutrino and $L_e = -1$ for the antiparticles (positron and antineutrino). This number is typically conserved in reactions, such as in beta decay, where $L_e = 0$ before the decay and remains zero when an electron and antineutrino are formed in the decay, with $L_e = 1 + (-1) = 0$. There are analogous lepton numbers L_μ for the muon family and L_τ for the tau family. Although all three lepton numbers are conserved in most reactions, this is violated in neutrino oscillations, where a neutrino changes from one family to another.

WHAT IS THE HIGGS BOSON?

One of the more puzzling questions in physics is why the different particles have the masses they do. In the 1960s British physicist Peter Higgs developed a scheme for addressing that question. The idea is that each particle's mass is determined by its interaction with a field (now called the **Higgs field**) that exists everywhere in space. The **Higgs particle**—or **Higgs boson** because it has zero spin—is the mediator between the Higgs field and each particle in the field. The photon, which is massless, does not interact with the Higgs field. Low-mass neutrinos interact slightly with the field and other leptons and quarks even more, giving them more mass.

The Higgs boson is difficult to observe for two reasons. First, it is very short-lived, predicted by theory to have a mean lifetime on the order of 10^{-22} s. Second, it is more massive than even the W and Z bosons, with a mass of over 100 GeV/c^2. Therefore, it was not possible to observe the Higgs boson until the extreme energies of the LHC became available. Even then, the Higgs particle is not observed directly, but rather indirectly from the shower of other particles it produces when it decays. Observations of this kind in 2012 were deemed to have confirmed the existence of the Higgs boson.

WHAT IS UNIFICATION?

Physicists try to understand nature systematically and look for the simplest possible patterns. There are many examples in history of how this works. All the atoms in the periodic table are made up of just three particles—electrons, protons, and neutrons, with the protons and neutrons combined in a small nucleus. Electricity and magnetism, which seem to have different sources and manifestations, were unified into electromagnetism in the nineteenth century. In the twentieth century, the vast particle zoo was discovered, but we now think of those many particles as composed of different arrangements of just six quarks and six leptons (or twelve of each, counting antiparticles).

The unification of electromagnetism leaves four fundamental forces to consider: electromagnetism, gravitation, and the strong and weak interactions. Each of these interactions appears to be mediated by a boson or bosons: photons for electromagnetism, gluons for the strong force, and W and Z particles for the weak interaction. Another boson, the graviton, is believed to mediate the gravitational force, but it hasn't been observed.

In the 1960s Abdus Salam, Sheldon Glashow, and Steven Weinberg developed a theory that successfully unified electromagnetism and the weak interaction into a single electroweak interaction. Since then there have been many attempts at a grand unified theory, which would combine the strong force with the electroweak. There have been no successful grand unified theories, but some have contained interesting predictions such as proton decay and magnetic monopoles, neither of which has been observed.

Ultimately, physicists not only would like to find a grand unified theory but also include gravity in a "theory of everything" that would explain all four fundamental forces. Gravity seems very different from the other interactions, and it's nearly 40 orders of magnitude weaker than electromagnetism. There is plenty of theoretical work left to be done!

FURTHER READINGS

Feynman, Richard P., and Weinberg, Stephen 1987. *Elementary Particles and the Laws of Physics: The 1986 Dirac Memorial Lectures.* Cambridge, England: Cambridge University Press.

Griffiths, David 2008. *Introduction to Elementary Particles.* Weinheim, Germany: John Wiley-VCH.

Thornton, Stephen T., and Rex, Andrew 2013. *Modern Physics for Scientists and Engineers,* 4th ed. Boston: Cengage Learning.

Chapter 9

Relativity

Relativity is one of the most important and fascinating fields of study within physics. It's generally split into two subfields—special relativity and general relativity—both of which were developed by Albert Einstein in the early twentieth century. Relativity accounts for how events are viewed from different frames of reference, when those frames of reference are moving with respect to one another or in strong gravitational fields. It also accounts for how light will appear to observers in different frames of reference. Many of the effects predicted by the theory of relativity are contrary to the predictions of classical physics and to perceptions based on everyday experience. These striking and sometimes shocking predictions have been verified repeatedly by experimental evidence. Probably the most famous prediction of special relativity is the equivalence of mass and energy, expressed in the formula $E = mc^2$. This equivalence manifests itself in many ways, particularly in atomic and nuclear physics (Chapter 7) and in the study of fundamental particles (Chapter 8). The implications are far reaching and include the production of nuclear energy and nuclear weapons.

WHAT'S THE DIFFERENCE BETWEEN SPECIAL RELATIVITY AND GENERAL RELATIVITY?

Special relativity accounts for how material objects and light are viewed by observers traveling in different frames of reference, in constant, nonaccelerated motion with respect to one another. The fact that the speed of light is the same for observers in all reference frames is of fundamental importance. It leads to correct predictions about how the different observers measure the positions, lengths, and velocities of objects, and how they measure the passage of time. The theory of special relativity also includes the equivalence of mass and energy.

General relativity extends the principles of special relativity to frames of reference that are in accelerated motion with respect to one another or, equivalently, in the presence of a gravitational field. The theory of general relativity not only accounts for the motion of material bodies through gravitational fields but also describes how light is affected by gravity. One of the well-known predictions of general relativity is that light moves in a curved path through a gravitational field. Another related prediction is the existence of black holes—objects so dense that they allow no light to escape.

HOW WAS SPECIAL RELATIVITY DEVELOPED?

By the beginning of the twentieth century, physicists had made a good deal of progress toward understanding the properties of light. They understood light to be an electromagnetic wave, with its measured speed in vacuum (about 3.00×10^8 m/s) predicted correctly by Maxwell's electromagnetic theory. In other media, light was found to travel more slowly, consistent with the refraction that takes place when it crosses a boundary between different media.

There remained the question of how observations of light are affected by the relative motion of observers. In 1727 English astronomer James Bradley discovered **stellar aberration,** which is the apparent motion of stars related to the motion of Earth in its orbit around the sun. (This is distinct from **stellar parallax,** the apparent change in stars' positions due to differences in Earth's position throughout the year.) Bradley explained aberration by assuming that light is made up of a stream of particles. He reasoned that catching the light in a telescope is analogous to catching a stream of falling raindrops. If the raindrops are falling straight down, then to an observer moving horizontally, their velocity would have a horizontal component opposite to that of the observer. The observed aberration seems to confirm this. However, by the early nineteenth century, evidence seemed to favor the idea that light is a wave, not a particle. A wave should require a medium of transport, and that medium was dubbed the **luminiferous** (light-carrying) **ether.** The observation of stellar aberration required that Earth move through the ether, which would be at rest relative to the sun.

In the 1880s, American physicists Albert Michelson and Edward Morley attempted a direct measurement of the effects of relative motion by observing the interference pattern between two beams of light traveling perpendicularly to one another. They built their apparatus so that they could vary the directions of the light paths with respect to Earth's orbital motion through the supposed ether. They famously obtained a null result—meaning that they could observe no change in the interference pattern, regardless of the light paths. The null result seemed to contradict the observations of stellar aberration, at least as it was interpreted in the ether-based model.

In the early 1900s Einstein thought seriously about the issue of how light propagates and how it appears to different observers. Because light is an electromagnetic wave, a related question is how electric and magnetic fields appear to different observers, and Einstein also considered this question. For example, an electric charge at rest creates an electric field but no magnetic field. However, to a moving observer, that same charge is moving and therefore generates a magnetic field in addition to an electric field. A full treatment of relativity would be needed to reconcile these observations. In 1905 Einstein provided a complete theory of special relativity, which accounted for the behavior of light and electromagnetism.

WHAT ARE THE TWO PRINCIPLES (OR POSTULATES) OF SPECIAL RELATIVITY?

Einstein thought that it was unreasonable to assign a state of absolute rest to the ether with respect to our sun. He went further by saying that no place could be considered at absolute rest or privileged in any other way. That essentially was his first principle of relativity. His second principle, consistent with the first, made the special point that the speed of light should be the same for different observers. Strictly speaking, these principles apply to inertial (non-accelerated) frames of reference, in which Newton's law of inertia holds. To summarize, the two principles of relativity are

1. There are no preferred frames of reference, and the laws of physics appear the same in all inertial frames.
2. The speed of light in a vacuum (c) is the same for observers in all inertial frames.

On the surface these principles might seem sensible enough, even obvious. But if you think about them closely, you'll see that they are at odds with some aspects of everyday experience. For example, if you're riding a bicycle at 10 mph and throw a ball in the same direction at 20 mph relative to you, then it's moving 30 mph with respect to the ground—a different speed. Yet according to Einstein's postulates, if you shine a flashlight beam from your moving bicycle, both you and an observer on the ground will measure the light's speed to be the same—namely, c. These two simple principles have enormous consequences, not only for light but also for all distance and time measurements.

HOW ARE TIME AND DISTANCE MEASUREMENTS AFFECTED IN RELATIVITY?

One of the striking results of the principles of relativity is the effect on the measurement of time. Suppose you are in an inertial frame and measure a time interval Δt. The same time interval measured in a second inertial frame, moving at speed v relative to the first, is $\Delta t/\gamma$, where γ is the **relativistic factor**

$$\gamma = \frac{1}{\sqrt{1 - v^2/c^2}} \tag{9.1}$$

and c is the speed of light in vacuum. This effect is known as **time dilation.** The dependence of the factor γ on speed v is shown in Table 9.1. At low speeds, or even at speeds that are high by everyday standards but small compared to the speed of light, γ is very close to 1, so there is little noticeable effect. However, the effect of time dilation becomes more pronounced the closer the relative speed gets to c. Time dilation is an established fact experimentally. It is easy to observe in fast-moving subatomic particles, where the effect is most pronounced, but now it can be observed routinely using precise atomic clocks traveling at much slower speeds.

Measurements of distances are similarly affected by relative velocities. Suppose you have an object of length L at rest in your inertial frame. An observer in a second inertial frame, moving with speed v in a direction parallel to the object at rest in your frame, measures the object's length to be L/γ, where again γ is given by Equation (9.1) and Table 9.1. This effect is called **length contraction.** The fact that length contraction and time dilation are governed by the same factor γ is not a coincidence but rather a result of the fact that the speed (distance/time) of light c is the same for all inertial frames, by the second principle of relativity.

Notice that the relativistic factor γ is always greater than 1. This means that a moving clock always runs slow, relative to a stationary one. Similarly, a moving object appears to be contracted (shorter) in length, in the direction of motion.

Only massless particles (such as photons) may travel at the speed of light c. Particles with mass must travel more slowly than c. Equation (9.1) shows that with $v > c$ the relativistic factor γ would be the square root of a negative number and therefore imaginary, not real. This is consistent with the prohibition on speeds greater than c. For an object in motion at the speed of light ($v = c$), the relativistic factor γ is undefined, consistent with the fact that an object with mass cannot reach the speed of light.

What Are Proper Time and Proper Length?

For the novice, it can be difficult to interpret the effects of time dilation and length contraction because they seem to defy common sense. For example,

TABLE 9.1   THE RELATIVISTIC FACTOR Γ

Speed v (m/s)	Object Description or Fraction of Speed of Light	Relativistic Factor γ
30	Car at highway speed	1.000000000000005
250	Commercial jet	1.0000000000003
8,000	Orbiting satellite	1.00000000036
30,000	Earth's orbital speed 0.0001c	1.000000005
3×10^6	0.01c	1.00005
3×10^7	0.10c	1.005
1.5×10^8	0.50c	1.15
2.7×10^8	0.90c	2.30
2.85×10^8	0.95c	3.22
2.97×10^8	0.99c	7.34

suppose observers in inertial frames A and B each have a clock at rest in their own frame. The observer in A will find that B's clock runs slow, relative to the clock in A. However, because there is no preferred frame of rest, the observer in B finds that A's clock runs slow relative to the one in B. Common sense seems to tell us that either clock A or B should be slower in an absolute sense, or else they read the same. But relativity gives a different result because the start and end of a measured time interval may occur at different places in the two reference frames. Similarly, a length measurement depends on the simultaneous measurement of the two ends of an object, and two events that are simultaneous in one inertial frame are not simultaneous in another.

The concept of **proper time** helps avoid confusion and contradiction. A proper time interval between two events is measured at the same place in an inertial frame. The time interval between those same events seen in another inertial frame is shorter than the proper time by the factor γ. **Proper length** is the length of an object measured at rest in an inertial frame. The same length measured in another inertial frame is shorter than the proper length by the factor γ.

How Are Time Dilation and Length Contraction Related?

A classic experiment that shows the connection between time dilation and length contraction was first performed in 1940 by Bruno Rossi and D. B. Hall.

The experiment uses muons, which are created in Earth's upper atmosphere and travel toward Earth's surface at speeds of about $0.995c$. At such high speeds the relativistic factor γ is close to 10. Muons decay with a mean life-time of only 2.2×10^{-6} s. Classically, a muon can travel less than 1 km during that time, and therefore almost all the muons created in this way should decay before reaching Earth's surface.

However, a significant number of muons are detected at Earth's surface, and special relativity explains why. From the viewpoint of an Earth-based observer, the rapidly moving muon's internal clock is running slow by a factor of γ. The number of muons detected on Earth matches the prediction of special relativity.

Because all inertial frames are equivalent, the situation must also be considered from the muon's rest frame. In that frame, its clock is running normally, but the distance from the upper atmosphere to Earth's surface is contracted by the same factor γ. The number of muons that can safely traverse this shorter distance matches the number predicted from Earth's rest frame using time dilation. Experimental results predicted from the different rest-frame viewpoints must agree, and they do.

What Is the Twin Paradox?

The **twin paradox** is a hypothetical situation used to illustrate the effects of time dilation. It uses a pair of twins (call them Alice and Ben), so that the two people begin with the same age. Then Alice stays home while Ben travels at a constant speed of $0.87c$ relative to Earth until he reaches a nearby star system. The speed $0.87c$ is convenient because it makes $\gamma = 2.0$. After spending a negligible time at the other star system, Ben turns around and returns home at the same constant speed $0.87c$.

How do the ages of the twins compare when they are reunited upon Ben's return? Suppose the round trip took 30 years according to Alice's clock on Earth. With a time dilation factor 2.0, Ben should have aged only 15 years and will appear considerably younger than Alice. That's the correct interpretation. The nature of the paradox is that you may be tempted to consider the passage of time from Ben's view. Can't he say that Alice was moving at $0.87c$ relative to him, and thus shouldn't he find her younger than he is when they reunite? This second interpretation is invalid under the requirements of special relativity because Ben must undergo considerable acceleration in order to turn around and return. Thus, he does not occupy the same inertial frame throughout the experiment. Alice does remain in the same inertial frame (ignoring the slight noninertial motion she experiences on a rotating, sun-orbiting Earth), so her view is the correct one.

Can You Travel Backward in Time?

The time dilation effect in special relativity slows the passage of time in one reference frame relative to another, as shown by the twin paradox. However, traveling backward in time is not permitted. One clock may tick very slowly compared with another. For example, the 7-TeV protons generated by the Large Hadron Collider particle accelerator (LHC, Chapter 8) have a relativistic factor γ of over 7,000, meaning that their internal clocks run slow by that factor, relative to a clock fixed in the lab. Further acceleration will only increase γ but will never reverse the flow of time.

WHAT IS SPACETIME?

The concept of **spacetime** combines the dimensions of space and time into a single multidimensional space and in doing so exemplifies the close connections between space and time in special relativity. For one-dimensional motion along the *x*-axis, spacetime has two coordinates, *x* and *ct*. The time coordinate is represented as *ct* rather than *t* so that it will have the same dimensions and units as the spatial coordinate(s).

Figure 9.1 shows a common graphical representation—a **spacetime diagram**—for motion in one dimension (*x*). Space is represented on the

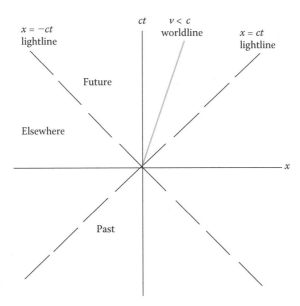

Figure 9.1   Features of a spacetime diagram for one-dimensional motion.

horizontal axis and time on the vertical. The "lightlines" $x = ct$ and $x = -ct$ are the paths of light moving at speed c sent in both directions from the origin ($x = ct = 0$). The "worldline" of a particle or object tracks its position and time. The worldline of any massive object that starts from the origin must lie between the lightlines because it's constrained to have speed less than c. The region between the lightlines with $ct > 0$ is called the future because it occurs after $t = 0$; similarly, the region between the lightlines with $ct < 0$ is the past. The part of spacetime outside the lightlines is called "elsewhere" because it's inaccessible to an object whose worldline passes through the origin.

For two-dimensional motion there is (not shown) a second spatial dimension, perpendicular to the x and ct axes. The lightline rotates around the ct axis to form a light cone. For three-dimensional motion, a fourth dimension is needed to represent the three spatial dimensions plus time, but this is difficult to visualize.

One advantage of spacetime is that it allows for the definition of a **spacetime interval.** For two events in one dimension separated in space by Δx and in time by Δt, the spacetime interval is

$$\Delta s^2 = \Delta x^2 - c^2 \Delta t^2 \qquad (9.2)$$

The spacetime interval is the same when measured in different inertial frames. Although observers in different inertial frames will measure different values for Δx and Δt, the quantity $\Delta s^2 = \Delta x^2 - c^2 \Delta t^2$ is the same for all.

HOW ARE VELOCITY MEASUREMENTS AFFECTED BY HIGH SPEEDS?

What happens when one velocity is added to another? An example of this is if you are traveling in a car at 10 mph and throw a ball at 20 mph relative to you in the forward direction. Relative to the ground, the ball's speed is 10 mph + 20 mph = 30 mph. There's nothing wrong with adding velocities this way when speeds are low, but at high speed there's an obvious problem. Consider what would happen if you were traveling away from Earth in a spaceship at half the speed of light (0.5c) and fired a beam of electrons in the forward direction at 0.9c. Following the example of the car and ball, the electrons' speed relative to Earth would be 0.5c + 0.9c = 1.4c. This is clearly incorrect because the laws of relativity prohibit any measured speed greater than c.

Following a procedure similar to the one used to find the relativistic factor γ for time dilation and length contraction, a correct relativistic formula can be found. Consider two inertial systems, with one considered fixed and another having velocity v_1 with respect to the fixed system. If an object has velocity v_2

with respect to the moving system, then that object's velocity with respect to the fixed system is

$$v_{rel} = \frac{v_1 + v_2}{1 + v_1 v_2/c^2}$$ (9.3)

For the spaceship and electron beam, with $v_1 = 0.5c$ and $v_2 = 0.9c$, the result is $v_{rel} = 0.966c$, which is very fast but safely under the universal speed limit c. The structure of the formula ensures that the result is always less than c, no matter how close the individual velocities are to it. For low speeds, the denominator in Equation (9.3) is very close to 1, and the formula reduces to approximately $v_{rel} = v_1 + v_2$, in agreement with the classical and commonsense result.

The velocity addition formula is consistent with the speed of light being constant in all inertial frames. Suppose the spaceship traveling at $v_1 = 0.5c$ away from Earth shines a beam of light ($v_2 = c$) in the forward direction. Then $v_{rel} = (0.5c + c)/(1 + 0.5) = c$, so the light beam is traveling at speed c relative to Earth as well. If, instead, the spaceship sends a beam of light back toward Earth ($v_2 = -c$), then $v_{rel} = (0.5c - c)/(1 - 0.5) = -c$, which is a speed of c but moving toward Earth, as it should be.

There are comparable velocity addition formulas for motion in two and three dimensions, but they are too complex to present here. Like the one-dimensional formula in Equation (9.3), they preserve the speed of light in all inertial frames and approach the classical results in the case of low-speed motion.

WHAT IS THE DOPPLER EFFECT?

The **Doppler effect** is the change in frequency or wavelength due to the relative motion of a source and receiver of waves. Most people are familiar with the Doppler effect for sound waves (Chapter 3), based on the sounds they hear from large vehicles such as trucks and trains. If such a vehicle is moving toward you and then passes and starts moving away, you can notice a reduction in the frequency (or pitch) of the vehicle sounds that reach you. In general, relative motion of any wave source and receiver toward one another results in a higher frequency of receipt, relative to the case when both source and receiver are stationary. Because wavelength and frequency are inversely related, higher frequency corresponds to shorter wavelength. When the relative motion of wave source and receiver is away from one another, then a lower frequency (longer wavelength) is received.

The Doppler effect also works for light and other electromagnetic waves, which all travel with the same speed c in vacuum. When a source and receiver approach with relative speed v, the wavelength received is

$$\lambda = \lambda_0 \sqrt{\frac{1 - v/c}{1 + v/c}} \tag{9.4}$$

where λ_0 is the source wavelength. For example, if a 650-nm red light source moves toward a receiver at a speed of $0.1c$, the received wavelength is 588 nm, which is no longer red but yellow! If the source and receiver move apart with relative speed v, the wavelength received is

$$\lambda = \lambda_0 \sqrt{\frac{1 + v/c}{1 - v/c}} \tag{9.5}$$

If the same 650-nm source moves away from the receiver at $0.1c$, the received wavelength is 719 nm, which is past red and into the infrared part of the electromagnetic spectrum. Wavelength and frequency vary inversely with respect to one another, so received wave frequencies increase when source and receiver are approaching and decrease when they move apart.

The Doppler shift toward shorter wavelengths—generically, the blue end of the visible spectrum—is called a **blueshift.** Similarly, a Doppler shift toward longer wavelengths is called a **redshift.** Observations of blueshifts and redshifts are important in astrophysics because they reveal the motion of light sources (such as stars and galaxies) relative to Earth. The observed spectra of all distant galaxies exhibit redshifts, and the redshifts become more extreme as our distance from the galaxy increases. The redshifted spectra from distant galaxies provide strong evidence of an expanding universe.

 WHAT IS RELATIVISTIC ENERGY?

Like time, distance, and velocity, computations involving energy work differently in special relativity than in classical mechanics. Classically, a particle with mass m and speed v has kinetic energy $\frac{1}{2}\, mv^2$. This obviously can't work for high particle speeds. The speed limit c for particles would then imply a maximum value $\frac{1}{2}\, mc^2$ for kinetic energy. However, a particle already near the speed of light can continue to have work done on it, increasing its kinetic energy without bound. The correct relativistic expression for kinetic energy is

$$K = mc^2(\gamma - 1) \tag{9.6}$$

where γ is the same relativistic factor given in Equation (9.1). For low speeds, Equation (9.6) approaches the classical formula $K = \frac{1}{2}\, mv^2$. Note that $\gamma = 1$ when $v = 0$, so the kinetic energy is zero for a particle at rest, as you'd expect. For very high speeds (approaching c), the factor γ continues to grow with increasing speed, and there is no limit to a particle's kinetic energy.

One of the more striking predictions of special relativity is that a particle of mass m at rest has an equivalent energy

$$E_0 = mc^2 \tag{9.7}$$

E_0 is called the **rest energy.** Equation (9.7) represents a profound fact of nature: Mass and energy are completely interchangeable. Mass can be transformed into energy and vice versa. In many nuclear reactions (including alpha and beta decay, fission, and fusion), the net mass of all the particles after the reaction is less than the net mass before the reaction. The lost mass Δm shows up in energy equal to Δmc^2, often as kinetic energy in the recoiling particles. This is the source of the energy produced in nuclear reactors and nuclear weapons. Conversely, other forms of energy can be converted into mass. An example of this is in a particle accelerator, where subatomic particles are accelerated to very high kinetic energies. When high-energy particles strike a fixed target or other particles in a collision, the kinetic energy they lose is used to form new particles. This is the source of the many new particles physicists have discovered since the early twentieth century (Chapter 8).

Physicists always look for conserved quantities in order to better understandw physical processes. When any reaction occurs that involves the transformation of mass and energy, the total **mass–energy** of the system is conserved. That is, any loss of mass is accounted for by an increase in energy, and vice versa. Mass–energy is one of the fundamental conserved quantities in nature. Other important quantities that are always conserved include momentum, angular momentum, and electric charge.

It takes a considerable amount of energy to create a new particle. Even the lightest common particle, the electron, has a rest energy of 511 keV. The proton and neutron have rest energies of 938 and 940 MeV, respectively—nearly 2,000 times more than the electron. By the equivalence of mass and energy, a particle's mass in terms of its rest energy is $m = E_0/c^2$. For this reason, physicists often express particle masses in units of eV/c^2—for example, 938 MeV/c^2 for the proton.

For a free particle—not subject to outside forces so that potential energy can be neglected—its total energy E is the sum of kinetic energy and rest energy, or $E = K + E_0$. The total energy is also given by

$$E = \gamma mc^2 \tag{9.8}$$

This makes sense because, for a particle at rest, $\gamma = 1$, and the total energy equals the rest energy. For a moving particle, the total energy increases in proportion to γ.

What Are Photons?

Photons are the massless particles that carry electromagnetic energy. They travel at the speed of light, which requires them to be massless. (By Equation 9.8, a particle with mass traveling at the speed of light would have an infinite amount of energy.) Thus, you can think of photons as consisting of pure energy and zero mass.

GOING DEEPER—INVARIANT QUANTITIES IN RELATIVITY

Momentum is rigorously conserved for all systems, but at relativistic speeds the classical expression $p = mv$ is invalid. Just as with kinetic energy, a particle's momentum is not bounded at mc by the universal speed limit c. Relativistic momentum includes the relativistic factor γ:

$$p = \gamma mv \tag{9.9}$$

This expression approaches the classical result at low speeds, where γ is close to 1, but momentum can grow without bound at high speeds approaching c.

A particle's mass m, momentum p, and total energy E are related by

$$E^2 = p^2 c^2 + \left(mc^2\right)^2 \tag{9.10}$$

Rearranging this expression shows that the particle's rest energy is

$$E_0 = mc^2 = \sqrt{E^2 - p^2 c^2} \tag{9.11}$$

The particle's mass and rest energy are invariant. They depend only on the type of particle and not on how fast it's moving. Although p and E increase with increasing speed, Equation (9.11) shows that the combination $\sqrt{E^2 - p^2 c^2}$ is invariant. This fact is sometimes useful—for example, in finding a particle's momentum if its total energy is known.

There are numerous conserved or invariant quantities in special relativity. For a system of particles, the total momentum and mass–energy are conserved. Electric charge is also conserved in a system of particles. In a high-energy reaction, new charged particles may be created, but the net charge of the system must remain the same after the reaction. For a single particle, the energy-momentum invariant (Equation 9.11) applies. Measurements of position and time in different inertial frames are related by the spacetime invariant (Equation 9.2). Finally, perhaps the most important invariant in relativity is the speed of light c, which is the same in any inertial frame. For a new student of relativity, it's a good idea to keep all these invariants in mind, particularly when so many actual observations appear counterintuitive.

HOW IS RELATIVITY CONNECTED TO ELECTROMAGNETISM?

Maxwell's electromagnetic theory predicts the existence of electromagnetic waves that travel with speed c in vacuum, independently of frequency or wavelength (Chapter 2). This agrees with observations across the electromagnetic spectrum, from the lowest frequency radio waves to the highest frequency gamma rays. That's a frequency range of more than 20 orders of magnitude!

Einstein thought carefully about electromagnetism when he developed the theory of special relativity. Electric and magnetic fields must appear differently to observers in different inertial frames, due to relative motion of the source charges. Einstein reasoned that the correct transformations of space and time would be consistent with Maxwell's equations, which make correct predictions of electromagnetic effects. The rules Einstein developed for transforming from one inertial frame to another do give the correct results. The observed effects—for example, the motion of a charged particle in an electromagnetic field—reconcile perfectly when special relativity is taken into account.

WHAT IS THE PRINCIPLE OF EQUIVALENCE?

General relativity deals with accelerated (noninertial) reference frames and gravitational fields. At first these two situations might seem to have little to do with one another, but in fact they are closely related. If you are at rest in a uniform gravitational field, then any object subject to the field will feel a force and accelerate. A common example of this is dropping an object as you watch it in Earth's nearly uniform gravitational field. Now consider what happens if you are outside gravitational and other fields but accelerate uniformly. Relative to you, an object at rest (or in an inertial frame) will accelerate uniformly, just as the falling object near Earth. The **principle of equivalence** says that those two situations are in fact equivalent, meaning that there's no experimental test that can distinguish between them.

A related fact is that there are two kinds of mass. There's **inertial mass,** which is the ratio of the net force on an object to its acceleration, per Newton's second law of motion. Then there is **gravitational mass,** which determines the force of gravity when two bodies with mass attract one another. By the principle of equivalence, those two masses are equal. If they were not, an experiment could establish a difference between the uniform gravitational field and the uniformly accelerated frame of reference. Recent experiments have shown that inertial mass and gravitational mass are equivalent to within one part in 10^{12}.

HOW DOES GENERAL RELATIVITY DESCRIBE GRAVITY?

The classical (Newtonian) theory of gravity describes it as mutually attractive force between two masses, with the strength of the force proportional to both masses and the inverse square of the distance between them. Newtonian gravity can also be described by a vector field surrounding any object with mass. Each point in the field is a vector representing the attractive force toward the attracting object on a test object placed in the field. This classical model works well in many situations involving large bodies, such as the sun and planets, where it generally accounts for the orbits of planets and satellites. Newtonian gravity also gives a good description of the motion of falling bodies and projectiles near Earth's surface.

In general relativity, an object with mass affects the shape of the four-dimensional spacetime around it. Geometrically, the spacetime is flat when there is no mass present, and the presence of mass introduces curvature, which in turn affects other bodies with mass in that space. Figures 9.2 and 9.3 show a common two-dimensional representation of how this works. First, the presence of a massive body causes the spacetime around it to curve (Figure 9.2), analogous to how a heavy ball placed on a mattress causes an indentation. Figure 9.3 shows how a second object is affected by the spacetime curvature. The small body orbits the larger one, just like a satellite orbiting Earth. Escaping from the region of curved spacetime, such as when a rocket escapes Earth's gravity, requires more energy. The exact mathematical description of this effect, derived by Einstein, is rather advanced and won't be presented here.

WHAT ARE SOME CONSEQUENCES
OF GENERAL RELATIVITY?

One of the more striking predictions of general relativity is that light is affected by gravity. One way to think of this is that light simply follows a curved spacetime path around a massive object. You can also use the principle of equivalence to convince yourself that gravity must affect light. In empty space, light follows a straight path. If you accelerate perpendicularly to that path, then

Figure 9.2   Illustration of the warping of spacetime around a massive object.

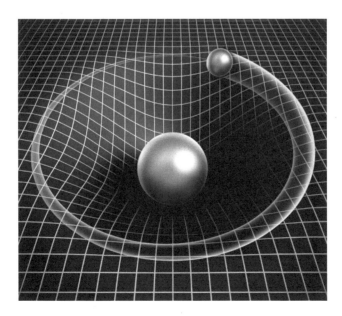

Figure 9.3   A smaller body follows the curvature of spacetime, resulting in a closed orbit.

the light's path is curved in your frame of reference. Now consider the situation at rest in a uniform gravitational field. By the equivalence principle, light traveling perpendicularly to the field must curve as it does in the accelerated frame of reference.

The effect of gravity on light is generally a small one, so it's best observed with massive bodies such as stars. The first observation of light curvature was made in 1919 during a total solar eclipse. The eclipse allowed stars near the sun to be photographed. Careful measurements of the stars' apparent positions revealed that the path of light from each star had curved slightly as it passed near the sun, by an amount in agreement with Einstein's general relativity theory.

The curvature of spacetime around massive bodies affects not only visible light but any kind of electromagnetic radiation. For example, radio signals sent back and forth between Earth and a spacecraft or planet on the other side of the sun follow a curved path. As a result it takes slightly more time for the signal to reach its destination, due to the longer path it must follow.

The elliptical orbits of planets around the sun are not fixed but precess slowly in time. The orbit's perihelion (closest approach to the sun) provides a convenient reference point, so this effect is called **perihelion precession.** Most of the effect is due to perturbations from other planets and can be accounted for by Newtonian gravity. However, for Mercury, the planet closest to the sun, some 43 arc seconds per century of perihelion precession cannot be accounted for classically. This was a great puzzle until Einstein showed that his general relativity theory explains the discrepancy.

Light (or any electromagnetic radiation) is affected by gravity even when it doesn't follow a curved path. For example, an electromagnetic signal sent straight up from Earth should lose energy due to the pull of gravity opposite its direction of motion. The energy of a photon is inversely proportional to its wavelength, so the upwardly directed photon's wavelength increases as its energy decreases. This is known as **gravitational redshift** and has been confirmed experimentally in different parts of the electromagnetic spectrum, with the measured redshift as predicted by general relativity. Similarly, a blueshift to shorter wavelengths occurs when light is directed downward in a gravitational field.

 ## WHAT IS A BLACK HOLE?

Light emerging from a planet or star suffers a small gravitational redshift. But could a denser object produce such a strong gravitational field that light has insufficient energy to escape? That's just what happens in a **black hole.** The boundary marking the point from which light can no longer escape is called the **event horizon.** General relativity theory predicts the existence of black holes, but they can be difficult to observe because (by definition!) they don't emit visible radiation. However, the intense gravitational field around a black hole can cause the formation of a so-called **accretion disc** around the black hole, which emits radiation that can be detected. Another method of black hole detection works when the black hole is part of a binary system, along with another star. Accretion of matter from the star into the black hole generates a strong x-ray emission, which in some cases signifies a black hole. Finally, it has been observed that light from the luminous star in a black hole–star binary is Doppler shifted due to the star's orbital motion around the binary system's center of mass.

It was suggested in the 1930s that black holes can be produced when a massive star collapses at the end of its lifetime. For most if its lifetime, nuclear fusion in the star's core generates substantial outward pressure, which balances the inward gravitational pressure in a fairly stable equilibrium. Once fusion stops and the outward pressure vanishes, collapse into a black hole can occur if the star is sufficiently massive—at least several times more massive than our sun. It's now believed that many black holes have been formed this way. There is also strong evidence that many if not most galaxies—including our own Milky Way—have in their centers a supermassive black hole, with mass a million to a billion times our sun's mass.

WHAT ARE GRAVITATIONAL WAVES?

A **gravitational wave** is a variation in gravitational field produced by a strongly accelerated mass. Good candidates for producing gravitational waves are rapidly rotating systems containing massive bodies, such as black holes and neutron stars. Even the closest of these objects is far from Earth, making gravitational waves difficult to detect against the background of other gravitational fields. Physicists are still seeking confirmation of the existence of gravitational waves, but according to theory they should exist.

HOW DO WE USE RELATIVITY?

Most of the effects of special and general relativity aren't obvious, and some are counterintuitive. However, whether or not you think about it, your existence on Earth depends on mass–energy equivalence because it's the conversion of mass to energy that provides the sun's energy. Apart from keeping us warm, the sun is the indirect source of most other forms of energy we use, including fossil fuels and hydroelectric energy. Energy from nuclear reactors doesn't involve the sun but rather comes from direct conversion of mass to energy. Many other effects, such as time dilation, length contraction, and relativistic velocity addition, are too small to notice in everyday life. Most of the tools and machines we use work just fine under the laws of classical physics.

A newer technology that depends on relativity is the **global positioning system (GPS)**. GPS uses an array of satellites in fixed orbits around Earth. A GPS receiver on Earth uses signals from at least four of the satellites, containing information on the satellite's position and the exact time each signal was sent. A microprocessor in the detector analyzes that information, using the fact that each radio signal travels at the speed of light. By using four transmissions, this fixes the receiver's position on Earth with a high degree of precision.

The system depends on accuracy of the satellite clocks, which are affected by relativity. A satellite in orbit is moving rapidly with respect to a ground-based clock, so the satellite clocks lose about 7 μs per day due to special relativity. The general relativistic effect of the clock being in the gravitational field of a high orbit is in the opposite direction, making the clocks run faster by 45 μs per day. The net effect is that the satellite clocks run 38 μs per day faster than an Earth-based clock. That may not sound like much, but it's vitally important to this system. In 38 μs, light (or a radio wave) travels over 11 km! If you want the GPS to give your position within 10 m, the clocks must be precise to within the time it takes light to travel that distance, which is just 33 ns. For GPS to remain accurate, the clocks must be designed and built with the appropriate correction factors in mind.

 ## SO IS EVERYTHING REALLY RELATIVE?

From a physicist's perspective, the answer to this question is a qualified yes. Physics depends on measurement and, generally speaking, the physical properties we mention depend on the frame of reference used. Many of the most fundamental quantities in physics—including position, velocity, force, and energy—depend on the motion of the observer relative to whatever is being measured.

On the other hand, many important invariant quantities are independent of relative motion. These include the speed of light in vacuum, electric charge, and other particle properties such as baryon number, rest energy, proper time, and proper length. Ironically, the theory of relativity is largely based on invariants.

Relativity has precise meaning in physics, dealing with measurements made from different reference frames. After relativity was introduced to the general public through popular media, the idea of relative points of view was often introduced in nonscientific discussions. Some of these applications disturbed Einstein, the creator of the two relativity theories. For example, he dismissed extensions of his theories that undercut cultural norms, particularly when arguments from physics were used to justify moral relativism. Einstein said that he wished that, instead of relativity, he had given his work the name "theory of invariants" so that it wouldn't have been applied in that way outside physics.

FURTHER READINGS

French, A. P. 1968. *Special Relativity.* New York: W. W. Norton & Company.

Mermin, N. David 1989. *Space and Time in Special Relativity.* Prospect Heights, IL: Wavelend Press.

Mermin, N. David 2005. *It's About Time: Understanding Einstein's Relativity.* Princeton, NJ: Princeton University Press.

Ohanian, Hans C., and Ruffini, Remo 2013. *Gravitation and Spacetime,* 3rd ed. Cambridge, England: Cambridge University Press.

Taylor, Edwin F., and Wheeler, John Archibald 1966. *Spacetime Physics.* San Francisco, CA: W. H. Freeman and Company.

Thornton, Stephen T., and Rex, Andrew 2013. *Modern Physics for Scientists and Engineers,* 4th ed. Boston: Cengage Learning.

Chapter 10

Astrophysics and Cosmology

Physics has long been applied to study astronomical objects, first in our solar system and then throughout the universe. It was thinking about how the moon moves around Earth and how the planets move around the sun that led Newton to develop his law of gravity in the late seventeenth century. Newtonian gravitation gives a good account of the motion of large bodies in the universe, and where Newton's theory falls short, it has been improved upon by Einstein's theory of general relativity. Better observational techniques in the eighteenth, nineteenth, and twentieth centuries allowed astronomers and astrophysicists to expand their studies well beyond our solar system—to more distant stars and eventually to the entire visible universe. Along the way they discovered different kinds of stars and learned how stars are formed and then evolve over time. These studies have been informed by advances in physics, particularly since 1900, when better understanding of atoms, nuclei, and fundamental particles helped scientists understand processes throughout the universe. We now have a better understanding of the structure of the universe, but a number of puzzles remain. Today most of the uncertainty centers around the matter we can't see but can only infer from its effects on visible matter. Much remains to be learned about the expansion of the universe and its eventual fate!

WHAT ARE ASTRONOMY, ASTROPHYSICS, AND COSMOLOGY?

Ancient peoples around the world were fascinated by the motions of the heavens—the sun, moon, and stars. These were the early astronomers—people who studied the heavens in an effort to understand what was going on well beyond Earth. Along with **astronomy,** many cultures speculated about the related field of **cosmology,** which addresses the origins, history, and future of the universe. Without telescopes or other modern scientific tools, ancient people often rooted their cosmologies in the religion and philosophy of their own cultures. By doing so they sought to give deeper meaning to their existence.

Today cosmology relies on scientific methods. This includes **astrophysics,** where the theories and methods of the physicist are applied to astronomy. Almost every physics subfield is applicable in some way to astrophysics, including mechanics; electromagnetism; spectroscopy; atomic, nuclear, and particle physics; quantum mechanics; optics; and thermodynamics.

 HOW WAS OUR SOLAR SYSTEM FORMED?

About 4.6 billion years ago, a cloud of molecules and dust initially much larger than our present solar system condensed due to mutual gravitational attraction. Most prevalent were single hydrogen atoms, but the heavier elements in our solar system were also present in complex molecules and dust grains. Eventually, enough mass—again, mostly hydrogen—condensed at the center to form the sun. The cloud already possessed angular momentum, so the remaining mass continued to circulate around the central body. As it did so, gravitational attraction formed other bodies from it, ranging from small rocks to larger masses that would eventually become the planets and their moons.

This process led to the eight planets we recognize today, which can be divided into two groups. Closer to the sun are the four **rocky** (or "**terrestrial,**" meaning Earth-like) **planets:** Mercury, Venus, Earth, and Mars. They are relatively rich in heavier elements, including metals (particularly iron, the most plentiful heavier element) and silicates, which trap significant amounts of oxygen. Beyond the orbit of Mars are the four **gas giants:** Jupiter, Saturn, Uranus, and Neptune. In those planets solid icy cores serve as the attracting centers for large clouds of gases (mostly hydrogen) that form the bulk of those planets' volumes. The light gases make the densities of the gas giants quite low, compared with the terrestrial planets. Jupiter's overall density (just over 1300 kg/m^3) is only slightly greater than that of water (1000 kg/m^3), and Saturn's overall density is even less (about 700 kg/m^3)! By comparison, Earth's density is 5500 kg/m^3. Each of the four giant planets has a system of rings orbiting around it, consisting of dust and small chunks of ice and rock. Saturn's rings are most prominent, easily visible using a decent telescope or binoculars.

The reason for the two distinct groups of planets is that at the time the solar system was formed, the hot solar nebula prevented ices from forming all the way to Mars. Therefore, the terrestrial planets accreted out of rock and metal (mostly iron). Beyond Mars it was cold enough for ices—primarily H_2O, CO_2, CH_4, and NH_3—to form. That ice served as the cores of the four gas giants.

The six planets beyond Venus all have at least one natural satellite, or moon. These are smaller bodies orbiting the central planet, and there are more than 170 known moons in the solar system. The moons have enough orbital angular momentum to continue in orbit without falling into the planet. However, smaller moons in particular are subject to gravitational perturbations from other moons and planets and from larger asteroids that may pass by.

How Was the Moon Formed?

Earth's moon is larger relative to the size of its primary planet than any other moon in the solar system. The most common theory of its origin is that the moon was formed soon after Earth when a large body of some kind collided with Earth, causing the ejection of a large amount of matter that then formed the moon through mutual gravitation. (This is the "giant impact theory.") Although the question has not been settled conclusively, this theory is currently favored over several other candidate theories. These include (1) the possibility that the moon simply formed at the same time as Earth, from the primordial solar system; and a separate theory (2) that the moon came from elsewhere inside or outside the solar system and was captured by Earth's gravity.

However, theory (1) has difficulty explaining the difference in the average densities of Earth and the moon. Earth's overall density (about 5500 kg/m^3) results from the fact that it is a mixture of rock (average density of about 3000 kg/m^3) and iron (density of 7900 kg/m^3). The moon's average density is only 3300 kg/m^3, indicating nearly all rock with very little iron. If the moon formed in place near Earth, its density should be closer to that of Earth.

Theory (2) has trouble explaining another chemical clue to formation: the ratios of the stable isotopes of oxygen ^{16}O, ^{17}O, and ^{18}O. Isotopic analysis of meteorites shows that the oxygen ratios varied with location in the solar nebula. However, the ratios of $^{17}O/^{16}O$ and $^{18}O/^{16}O$ indicate that the moon and Earth formed near each other in space.

The giant impact theory explains these chemical and isotopic paradoxes in the following way. After the formation of Earth and the separation of the iron core (in the center) from the rocky mantle (above the core), any impact would eject rocky, iron-poor material into space, which could explain why the moon is iron poor and maintained the same oxygen isotope ratios as Earth.

WHAT ARE THE ORBITS OF THE PLANETS?

The study of planetary motion holds an important place in the history of science. The ancient Greeks noticed that the stars appear in a fixed pattern with respect to one another, but the planets move daily with respect to the background of stars. (The word "planet" is derived from the Greek word for "wanderer.") They believed that Earth is stationary and at the center of the heavens, which revolve daily around Earth. Because Earth actually rotates once a day and travels around the sun once a year, placing Earth at rest made it necessary for the Greeks to develop a fairly elaborate model to describe the motions they saw, particularly the motion of the planets, which in reality orbit the sun as Earth does. Despite the difficulty, their refined model provided a reasonably accurate description of how the sun, moon, and planets move as viewed from Earth, and that model persisted until the sixteenth century (C.E.).

During the European Renaissance there was renewed interest in the question, particularly after the Polish astronomer Nicholas Copernicus presented a workable sun-centered model of the solar system in 1543. In the early seventeenth century, the sun-centered model gained wide acceptance, thanks to the work of Galileo Galilei and Johannes Kepler. Galileo developed the astronomical telescope and used it to observe the four largest moons of Jupiter, the phases of Venus, and sunspots that move across the face of the sun. This evidence, along with some physical arguments made by Galileo, all pointed to a sun-centered system. (For example, he explained why objects dropped from a tower appear to fall straight down, rather than being left behind by a rotating Earth, as it was believed they would.) Kepler carefully analyzed the planets' orbits and showed that they have elliptical shapes, not compounded circles as in the ancient Greek model. The precision of Kepler's ellipses—including an ellipse for Earth's orbit around the sun—provided solid evidence for the new model. Later in the seventeenth century, Newton showed (Chapter 1) that mutual gravitational attraction accounts for the motions of planets, their satellites, and other objects in the solar system such as comets and asteroids. Newtonian mechanics predicts that the orbit of a small body around the sun must be one of the four conic sections—a circle, ellipse, parabola, or hyperbola—and that's just what is observed. Only a small correction is needed for general relativity, and this was provided by Einstein in the early twentieth century (Chapter 9).

The orbits of planets are nearly elliptical, with the sun at one focus of the ellipse. Small deviations from the elliptical shape arise due to perturbations from other planets, and the sun itself moves due to its attraction to the planets, particularly to the heavyweight Jupiter. Planetary orbits are nearly coplanar, all within 7° of Earth's orbital plane, which is called the **ecliptic** plane. The orbit of Earth's moon is inclined just over 5° from the ecliptic, but many of the other planets' satellites have orbits far from the ecliptic. All planets orbit the sun in the same direction—counterclockwise, if you view the solar system from above Earth's North Pole. The orbital period of each planet depends on the planet's distance to the sun. Kepler showed that the square of the orbital period is proportional to the cube of the planet's mean distance to the sun, so periods increase

TABLE 10.1   THE PLANETS AND THEIR ORBITS

Planet	Orbital Period (years)	Mean Distance to Sun (AU)	Eccentricity
Mercury	0.241	0.39	0.206
Venus	0.616	0.72	0.0068
Earth	1.00	1.00	0.0167
Mars	1.88	1.52	0.0934
Jupiter	11.9	5.20	0.0484
Saturn	29.5	9.56	0.0557
Uranus	84.1	19.2	0.0472
Neptune	165	30.1	0.0086

as you go farther out. The closest planet, Mercury, has the shortest period, just 88 days, while the most distant planet, Neptune, has a period of 165 years.

The shape of each planetary ellipse is described by a parameter called **eccentricity** (e), which can be defined as the ratio of the distance between the ellipse's two foci to the ellipse's major (long) axis. A low eccentricity means that the ellipse is less elongated, and the ellipse approaches a circle in the limit $e \to 0$. An elongated ellipse has a higher eccentricity, approaching (but not equal to) 1. All the planetary eccentricities are fairly low, particularly Earth with $e = 0.0167$, so Earth's distance from the sun does not vary much throughout the year. Table 10.1 gives the eccentricities along with each planet's orbital period and mean distance to the sun in astronomical units (AU), where 1 AU is defined as Earth's mean distance to the sun.

WHAT OTHER OBJECTS ARE IN THE SOLAR SYSTEM?

The eight planets, along with their moons and rings, are the most prominent citizens of the solar system, but there are many others. Next in size to the eight planets are **dwarf planets.** There is still dispute about what constitutes a dwarf planet, but they must not be satellites of other planets. It's also not clear how many objects will qualify as dwarf planets because most candidates lie beyond the orbit of Neptune and are difficult to observe. The five objects currently classified as dwarf planets are, in order of increasing distance from the sun: Ceres, Pluto, Haumea, Makemake, and Eris. Ceres was discovered in 1801 and lies in the asteroid belt between Mars and Jupiter. Pluto was discovered in 1930 and was classified as the ninth planet until 2006, when it was reclassified as a dwarf planet. Its highly eccentric orbit ($e = 0.25$) lies mostly beyond but partly inside the orbit of Neptune. The other three dwarf planets were discovered in the twenty-first century and lie farther out.

Asteroids are smaller than dwarf planets but also orbit the sun independently of the eight planets, for the most part. There are millions of asteroids, ranging in size from about 10 to 1000 m across. A large concentration of asteroids appears in the asteroid belt, which lies between Mars and Jupiter. The asteroid belt is thought to have evolved from bodies present in the early solar system that failed to condense into a single planet due to perturbations from nearby Jupiter. Asteroids are made up of a variety of materials. Some are rich in carbon, some in silicon, and others in metal. Some contain small amounts of water ice.

Comets are distinct from asteroids in that they contain not only rocky material and dust but also significant amounts of water, ammonia, carbon dioxide, and carbon monoxide—all in solid (frozen) form. Comets originate in the Kuiper belt or Oort cloud, both far beyond the orbit of Neptune. Some comets make it to the inner solar system, where the sun can illuminate their thin atmosphere of gas and dust, called a coma. This makes the main body of the comet appear as a large, fuzzy ball when viewed with the naked eye. The sun also vaporizes the frozen gases and, together with dust, those gases form the comet's tail, its most distinctive visible feature. Most comets have highly eccentric elliptical orbits, with long orbital periods, and therefore they spend relatively little time in the inner solar system. The most famous of these is Halley's Comet, which has a period of 75.3 years and will next pass inside Earth's orbit in 2061.

Meteoroids are smaller particles, up to 1 m across, and they may have either asteroid-like or comet-like composition. A meteoroid that enters Earth's atmosphere is heated by atmospheric friction, forming a stream of glowing particles called a **meteor.** Any fragments that strike Earth's surface are called **meteorites.** Meteor showers occur at regular times throughout the year when Earth's orbit takes it through a particularly high concentration of meteoroids, often comet debris. Meteors are beautiful to watch, but larger meteorites can be dangerous. Earth and other bodies in the solar system exhibit impact craters in varying sizes. The impact of a large object on Earth is believed to have caused the mass extinction of dinosaurs and many other species nearly 70 million years ago.

Why Isn't Pluto Considered a Planet?

Pluto was discovered in 1930, after an intense search triggered by observations of irregularities in Neptune's orbit. The discovery was fortuitous because later studies showed that the small discrepancies in Neptune's orbit can be accounted for in other ways, and Pluto is too small to affect Neptune's orbit significantly. After Pluto's moon Charon was discovered in 1978, its orbital dynamics allowed astrophysicists to estimate Pluto's mass to be less than one-fifth the mass of Earth's moon and barely 4% of the mass of the lightest planet, Mercury. Thus, based on mass alone, it's difficult to consider Pluto in the same class as the planets. Further, Pluto's orbit is more eccentric than that of any planet and also more inclined, with its orbital plane making a 17° angle with the ecliptic. After being named as a planet in 1930, Pluto was reclassified as a dwarf planet in 2006.

What Is the Solar Wind?

In the sun's outer layer (corona), gases heated to one million degrees K or more become ionized, and some of the hot nuclei and electrons are moving fast enough to escape the sun's gravity. This creates a stream of charged particles, called the **solar wind.** The solar wind consists mostly of hydrogen nuclei (protons) and electrons, with some helium nuclei (alpha particles) and heavier nuclei. Particles in the solar wind typically have kinetic energies ranging from 1 to 10 keV. When the solar wind reaches Earth, the strong magnetic field (especially near the poles) accelerates the charged particles in it, which subsequently release some of their energy as visible light in aurora displays.

 ## WHAT ARE STARS?

Stars are large, hot, luminous objects that pervade the universe. Our sun is an example, and it's but one of many billions of stars in the Milky Way galaxy. In turn, there are billions of galaxies in the universe. Accordingly, the number of stars in the universe is, well, astronomical.

Stars are formed when a sufficiently large mass of gas, almost entirely hydrogen, coalesces due to mutual gravitational attraction. Eventually, the core of the star becomes extremely dense and hot, which allows fusion of the hydrogen nuclei (protons) to take place. The interior of the star is so hot that electrons can't be bound to atoms, and instead the free nuclei and electrons together form plasma. The fusion process releases a large amount of energy by converting a portion of the initial mass into energy ($E = \Delta mc^2$, Chapter 9). A common process called the proton–proton chain eventually converts four protons into an alpha particle (helium nucleus), plus positrons and gamma radiation. Three helium nuclei can fuse into carbon and then, in the carbon cycle, carbon nuclei continue to react with protons to release more energy. Fusion of carbon and helium may continue until iron nuclei are formed—specifically ^{56}Fe because it has the largest binding energy per nucleon of any nuclide (Chapter 7). Fusion of iron with itself or with lighter nuclei would require energy input, so that is normally where fusion must stop.

As fusion proceeds throughout the main part of a star's lifetime, there exists equilibrium between the outward pressure from the core, created by the extreme amount of energy released, and inward pressure from gravitation, due to the star's extreme mass. Thus, our sun, for example, is relatively stable and will be for many years to come. However, eventually all the light elements will be fused as far as possible. Fusion ends, and the main part of the star's lifetime is over. The evolution and eventual fate of a star depend on its mass.

How Do Stars Evolve in Time?

Stars like our sun spend about 90% of their time undergoing hydrogen fusion to form helium. During this time they are said to be **main-sequence stars.** Their energy output grows slowly but steadily in time, as the mass of helium increases relative to hydrogen. Our sun is about halfway through its 10-billion-year lifetime, and its energy output has increased some 40% over the initial value. Roughly 5 billion years from now, when most of the hydrogen has been used, fusion in the core stops, causing gravitational contraction. This results in significant heating, which actually causes a faster rate of fusion of the remaining hydrogen, and the star expands to form a **red giant.** In the red giant phase, the sun will expand from its current radius (about 700,000 km) to nearly the radius of Earth's orbit (1.5×10^8 km). Due to rapid burning, the red giant phase may last only a few million years, after which the star sheds its outer core and becomes a **white dwarf.** A white dwarf is very dense but can no longer support fusion and will eventually cool to the point where it's no longer luminous. The maximum mass of a white dwarf is about 1.4 solar masses, where, by definition, the sun's mass is exactly one solar mass. Our sun will eventually become a red giant and then a white dwarf.

For larger stars, the greater gravitational pressure forces protons and electrons to combine, forming neutrons in an extremely dense object called a **neutron star.** There the neutrons are packed to roughly the density of a nucleus. That's many orders of magnitude denser than ordinary matter or a main-sequence star. A star larger than about three solar masses can pass through the neutron-star phase to become a black hole (Chapter 9).

How Are Stars Classified?

You may have noticed that stars don't all appear the same, even when you view them with the naked eye. Though mostly white or colorless, many stars have a distinctive red or blue hue. Apparent color is related to the star's surface temperature, just as for a blackbody, where by Wien's law (Chapter 4) the radiative output is shifted to shorter wavelengths as temperature increases. In turn, the surface temperature of a main-sequence star depends on its size, with larger stars becoming hotter.

This regular variation for main-sequence stars is illustrated on a **Hertzspring–Russell** (or **HR**) **diagram** (Figure 10.1). For reference, the sun is near the center of the main sequence, with luminosity defined as 1.0 and surface temperature about 5800 K. Red giants are off the main sequence. Because they are larger than main-sequence stars, they have higher luminosity than main-sequence stars of the same surface temperature and color. White dwarfs are smaller and thus lie below the main sequence.

WHAT IS A SUPERNOVA?

A **supernova** is an extremely energetic output that occurs in some stars at the end of their lifetime. During a supernova, which may last up to a few months, the star emits roughly as much energy as it did throughout its lifetime (i.e., over billions of years). There are two separate pathways that can trigger a supernova. First, a white dwarf may gain enough mass—perhaps through accretion from a companion star—to begin carbon fusion, triggering a "thermal runaway" that appears as the supernova. The second and more common initiator is the sudden gravitational collapse of a massive star late in its lifetime. This pathway is restricted to stars larger than about eight solar masses, so it can't happen to the sun.

"Nova" is the Latin word for new. Historically, supernovas were thought of as new stars because they appeared suddenly where no star had been visible before. In reality, the star was there before but was too dim to be seen before the tremendous increase in luminosity. There are reports of such events from various cultures over the past few thousand years, including one in 1054 that resulted in the Crab Nebula. A 1604 supernova was used by Kepler and others as evidence of a changing universe. In recent times, a supernova visible to the naked eye appeared in 1987. It was the first to be subject to measurements by modern scientific equipment. Among other things observed, a burst of neutrinos was received on Earth several hours before the light—an indication that neutrino emission accompanies the collapse of the large star's core.

Supernovas are important to us in that they are believed to produce most of the supply of heavy elements present in the universe. In a main-sequence star, there's insufficient energy to allow fusion to proceed past ^{56}Fe. However, the intense nuclear reactions in a supernova produce heavier elements. (Red giants are the only other possible source of heavy elements.) Your life depends mostly on elements lighter than iron—hydrogen, carbon, oxygen, nitrogen, sodium, potassium, and calcium. But look past iron in the periodic table, and you'll see

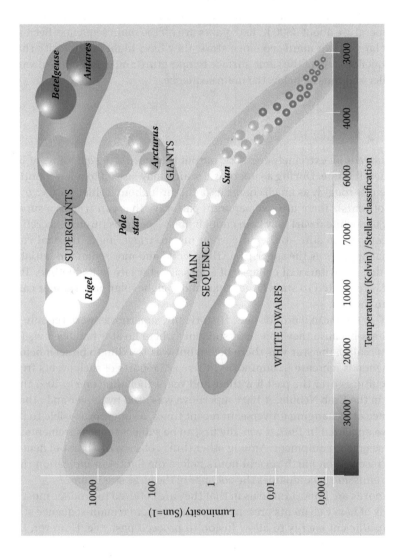

Figure 10.1   HR diagram showing different star classifications.

some other helpful elements, such as iodine, tin, silver, tungsten, platinum, gold, mercury, lead, and uranium. These all come from supernova explosions.

WHAT ARE GALAXIES?

A **galaxy** is a collection of a large number of stars, bound to each other gravitationally. Galaxies appear in many forms. We live in the Milky Way galaxy, which is believed to have a **spiral-arm** structure like the nearby Andromeda galaxy (a mere 2.5 million light years from Earth!), shown in Figure 10.2. Galaxies vary in size but typically have billions of stars. For example, the Milky Way is estimated to contain over 100 billion stars, and it's not extraordinarily large as galaxies go.

Visible in the Andromeda galaxy (Figure 10.2) is a large, bright core (or nucleus). The core is a common feature in galaxies. Physicists believe that many galactic cores contain a supermassive black hole, with masses ranging from millions to billions of solar masses. The shapes of spiral-arm galaxies are suggestive of rotation about the galactic core, similar to our solar system, and

Figure 10.2   The Andromeda galaxy.

that's just what happens. Although a galaxy's mass is more diffuse, relative to the solar system where most of the mass is in the sun, stars throughout the galaxy orbit around the center of mass. Observations of the stars' motions reveal important clues about the mass that's driving them, including not only visible matter but also gas, dust, and dark matter.

HOW HAS THE UNIVERSE EVOLVED IN TIME?

Physicists believe that the universe began in an event called the **Big Bang** approximately 13.8 billion years ago. At that moment, the entire universe was essentially contained within a point-like singularity, with density and temperature so high that they are difficult to estimate. Under those conditions, the four fundamental forces—gravity, electromagnetism, the weak force, and the strong force—were combined into a single force. This is the situation in which a theory of everything (Chapter 9) is required, but we currently lack such a theory.

Within the first fractions of a second after the Big Bang, the universe began to expand and cool. This caused the forces to separate into the ones we recognize. Gravity, the weakest force, was the first to become separate. Then the strong force became separate from the others. Quarks and leptons were formed, but the extreme temperatures ($>10^{16}$ K) kept them from joining together. This all happened within the first 10^{-13} s after the Big Bang! Shortly thereafter, the electromagnetic and weak interactions separated. With further cooling, quarks could join to form protons and neutrons. However, thermal energy kept protons and neutrons from joining to form nuclei until a few minutes later, when the universe had cooled to about 10^9 K. Only much later, when the universe had cooled to several thousand kelvins, could nuclei and electrons join to form neutral atoms.

Hydrogen was (and still is) the most common atom in the universe. However, the early universe was still too hot for significant amounts of hydrogen to coalesce gravitationally to form stars. That did not occur until perhaps 200 million years after the Big Bang. Star and galaxy formation could then proceed to form the universe as we see it today. Our sun is a relative latecomer in the universe, forming about 9 billion years after the Big Bang, or two-thirds of the way to the current age of the universe. Ever since the Big Bang, the universe has continued to expand and cool.

What Is the Evidence for the Big Bang and Expanding Universe?

You might well wonder how all this information about the Big Bang and early expansion of the universe is known, when there was no one around yet to observe it! People have long wondered about the origins and structure of the universe, but only within the last century or so have scientists had the tools needed to address these questions.

In the early twentieth century, American astronomer Edwin Hubble performed a painstaking survey of many galaxies spread throughout the universe. He noticed that the spectral lines of known elements were generally "redshifted"—that is, shifted to longer wavelengths. According to the Doppler effect (Chapter 9), a redshifted spectrum indicates that the source being observed has relative motion away from us. The fact that redshifts are observed for distant galaxies in all different directions indicates overall expansion of the universe.

Hubble and other scientists used the Doppler shift to measure each galaxy's relative velocity, and they compared each velocity to the galaxy's distance from us. There was a striking correlation—namely, a linear relationship between velocity and distance. This fact is known as **Hubble's law** and is expressed in the following simple equation:

$$v = HR \tag{10.1}$$

where v is the galaxy's recession velocity, R is the galaxy's distance from us, and H is called the **Hubble parameter.** The Hubble parameter can change in time. Its current value, H_0, which is called the **Hubble constant,** is the reciprocal of the age of the universe.

Another important piece of evidence for the Big Bang and expansion of the universe was discovered in the 1960s by Arno Penzias and Robert Wilson. While studying microwave radio transmission, they noticed that their microwave antenna received a faint but constant signal that emulated a blackbody radiation curve for a source at a temperature of about 2.73 K. That corresponds to a peak wavelength of about 1.06 mm. This background radiation is left over from the Big Bang or, more specifically, from the time after the Big Bang when hydrogen atoms could form, allowing photons to pass through the universe. That occurred when the universe's temperature was about 3000 K, so the shift to 2.73 K represents a Doppler-shift factor of about 1100.

How Are Cosmic Distances Determined?

The relative velocity of a distant star or galaxy can be measured reliably using the Doppler shift of its spectrum. However, the validity of Hubble's law depends on knowing the distance to each object, independently of its velocity. This is difficult to determine. When the distances grow to millions or even billions of light years, even an entire galaxy can be very dim, and there's no cosmic meter stick that you can use to measure such distances.

To solve the problem, scientists use a **distance ladder** that begins with shorter distances such as the astronomical unit, which can be measured accurately. Stellar parallax gives the distances to nearby stars. Measuring longer distances involves using comparison stars as what are called **standard candles.** This concept assumes that two stars of a particular class have about the same intrinsic brightness, regardless of where they are located. Then the brightness of a distant star is compared with the brightness of a similar star that's closer and has a better known distance. The brightness comparison yields the distance to the more distant, dimmer star. For more distant galaxies, the comparisons may be with individual stars (if they can be seen) or entire galaxies or clusters of galaxies. When available, supernovae are good distance indicators for other galaxies. Obviously, the results grow more uncertain as the distance grows. However, after many years of gathering and assessing data from distant sources, scientists have good confidence in their methods and results.

 ## WHAT ARE DARK MATTER AND DARK ENERGY?

The Big Bang initiated expansion of the universe, which continues to this day. But will it continue forever? Think of what happens when a rocket is launched straight up from Earth. It may rise a great distance before falling back to Earth under the influence of gravity. On the other hand, a rocket launched with sufficient speed will escape Earth and never return.

The concept of escape speed is a good analogy for understanding how the universe might evolve. Although the universe is still expanding, it's conceivable that mutual gravitation might slow and eventually stop the expansion. Then the mutual attraction would cause the universe to contract and approach the conditions of the early universe in what has been called a "big crunch." The extreme concentration of energy might cause a rebirth and another expansion—a "big bounce." On the other hand, a universe expanding at a fast enough rate will never stop expanding—a "big freeze"—because the universe would have low density and high entropy, with all the hydrogen that fuels stars used up. But what's fast enough? For the rocket leaving Earth, this is a straightforward problem: All you have to do is measure the rocket's speed and altitude, and Newtonian mechanics will tell you whether it escapes. It's a much harder problem to understand the future expansion of a universe that's spread over vast distances with mass that appears in many different forms, from stars and galaxies to grains of dust and free fundamental particles.

Thus, determining the universe's future becomes a problem of measuring its current state accurately enough and then extrapolating. Redshift data from distant galaxies give good information on the motion of various parts—analogous to the escaping rocket's speed. A major difficulty lies in determining how much mass there is because so much of the mass in the universe can't be seen.

There appears to be a significant amount of **dark matter,** which does not emit or absorb visible light or any other kind of electromagnetic radiation. There are several candidates for what constitutes dark matter, but the most likely one is **weakly interacting massive particles** (WIMPs), which are affected only by gravity and the weak force. The lack of electromagnetic interaction would explain why these particles can't be seen. Because it's not visible, the existence of dark matter can only be inferred by observing its gravitational interaction with luminous matter. That interaction is significant. Recent studies indicate that dark matter is at least five times more prevalent (by mass) than ordinary matter in the universe. That suggests that there may be enough matter in the universe to lead to contraction someday.

However, another recent discovery suggests just the opposite. Since the late 1990s, increasing evidence shows that the universe's rate of expansion is actually increasing, not decreasing as one would expect based on mutual gravitational attraction. This is an entirely new interaction that can't be explained using any combination of the four known forces, and the source of the new interaction is called **dark energy.** Dark energy is so prevalent that it's estimated to account for more than half of the mass–energy in the universe, with most of the rest consisting of dark matter. Ordinary matter appears to make up only about 5% of the universe!

WHERE DO WE GO FROM HERE?

Astronomers and astrophysicists are actively studying dark matter and dark energy, still relatively recent discoveries. There's a significant overlap between cosmology and the foundations of physics, particularly general relativity and fundamental particles. This is an exciting time to do research because so much is unknown in all these fields. We continue to wonder about what makes up the universe and the laws that govern it, hoping eventually to have a better sense of the future.

FURTHER READINGS

Pasachoff, Jay M., and Filippenko, Alex 2014. *The Cosmos: Astronomy in the New Millennium,* 4th ed. Cambridge, UK: Cambridge University Press.

Thornton, Stephen T., and Rex, Andrew 2013. *Modern Physics for Scientists and Engineers,* 4th ed. Boston: Cengage Learning.

Zeilik, Michael 2002. *Astronomy: The Evolving Universe,* 9th ed. Cambridge, UK: Cambridge University Press.

Zeilik, Michael, Gregory, Stephen A., and Smith, Elske V. 1992. *Introductory Astronomy and Astrophysics,* 3rd ed. Philadelphia: Saunders College Publishing.

Index